CERAMIC POWDER PREPARATION: A HANDBOOK

CERAMIC POWDER PREPARATION: A HANDBOOK

by

Dibyendu Ganguli

and

Minati Chatterjee

Central Glass and Ceramic Research Institute
Calcutta, India

KLUWER ACADEMIC PUBLISHERS
Boston / Dordrecht / London

Distributors for North America:
Kluwer Academic Publishers
101 Philip Drive
Assinippi Park
Norwell, Massachusetts 02061 USA

Distributors for all other countries:
Kluwer Academic Publishers Group
Distribution Centre
Post Office Box 322
3300 AH Dordrecht, THE NETHERLANDS

Library of Congress Cataloging-in-Publication Data

A C.I.P. Catalogue record for this book is available
from the Library of Congress.

Printed on acid-free paper.

Printed in the United States of America

Contents

Preface

The idea of writing this book came to us when, while working on a review on ceramic powder preparation, we discovered that there was apparently no single and compact volume which would suitably introduce the practical aspects of the subject to an interested beginner or an information-seeker. However, once at the job, we found that the task was not easy, because the topic could be looked at from significantly different angles, and the requirements of the reader would vary accordingly. We therefore chose model readers who would like to know why and when it was necessary to synthesize tailored powders, and what were the options from which to select for implementation or futher innovation. The writing of the book has been oriented accordingly.

In a dynamic area of activity like the one under discussion here, no review can claim to have presented a complete picture. While we have tried to assemble as much information as possible within the limitations of size of this volume, the sense of inadequacy remained. The usefulness of the endeavour, therefore, can only be left to the judgement of the readers.

D. Ganguli
M. Chatterjee

Acknowledgments

Our sincere thanks go to all individuals and organizations who have helped with photographs and drawings; they have been acknowledged at appropriate places in the text. In addition, Prof. K. C. Patil, Bangalore and Dr. R. Sivakumar, Thiruvananthapuram have specially helped with reprints of and other materials from their publications.

Many colleagues at the Institute helped us in various ways: the constant help of Dr. B. Siladitya deserves special mention. Ungrudging help also came from Drs. G. De, R.N. Basu, D. Basu, A. Sanyal and many others. The appreciative encouragement of Dr. C. Ganguly, Director is thankfully acknowledged. We also express our gratitude to the Indian Ceramic Society for readily agreeing to prepare the camera-ready manuscript. No words of praise would be sufficient for Mr. A.K. Banerjee, the Executive Secretary of the Society, whose constant support proved to be an asset.

Probably no book has ever been written without seriously affecting the family lives of the authors. One of us (D. Ganguli) would like to put on record a deep sense of gratitude towards his wife, Shibani, who had to spend many lonely hours through the months this book was being written. Similarly, M. Chatterjee acknowledges the constant encouragement and help of her sisters during the same period.

CERAMIC POWDER PREPARATION: A HANDBOOK

1

CERAMIC POWDERS AND THEIR APPLICATIONS

1.0 INTRODUCTION

The word "ceramics" generally represents inorganic, non-metallic objects which are prepared by heating, and are capable of withstanding high temperatures. While there is apparently no prescribed lower limit for the qualifier "high", for practical purposes a tentative lower boundary of about 1000°C seems in order, at or below which a ceramic object (and the substance it is made of) should not decompose, vaporize or melt. There is a large family of inorganic crystalline substances — synthetic or natural — which fall directly under this category. When in particulate form, they are popularly called ceramic powders. Ceramic powders are also quite commonly obtained via the thermal treatment of various hydrated crystalline and dry or hydrated amorphous particulates which do undergo some changes in structure (e.g. amorphous → crystalline) and composition (e.g. loss of water) in the process. However, they are also called ceramic powders, or more logically, precursor powders for ceramics.

Most of the commercial or research powders are used in the fabrication of bulk ceramics, including those with dimensional preferences, e.g. ceramic substrates by various processing techniques. Apart from this, ceramic powders are also used directly, i.e. without forming and consolidation into bulk shapes. Instances are pigments, abrasives, feedstock for thermal spray coatings etc. Depending upon the application, ceramic powders can vary in size from a few nm to a few hundreds of microns (when they are more often called "grains"). A requisite size fraction is obtained not only by the physical means, i.e. crushing-grinding-sieving, but more importantly, by a control on the parameters of synthesis. For achieving this control, it is necessary to understand the phenomena in operation during or after the formation of a powder, and the constitution of the powders thus formed. A brief introduction to the subject will be made below. The Chapter ends with a short section on the areas of application of (mostly synthetic) ceramic powders.

1.1 WHAT ARE CERAMIC POWDERS MADE OF ?

1.1.1 The Basic Units

All ceramic powders are constituted of individual *primary* particles or their agglomerates (Fig. 1.1). Primary particles, in turn, are composed of *crystallites*

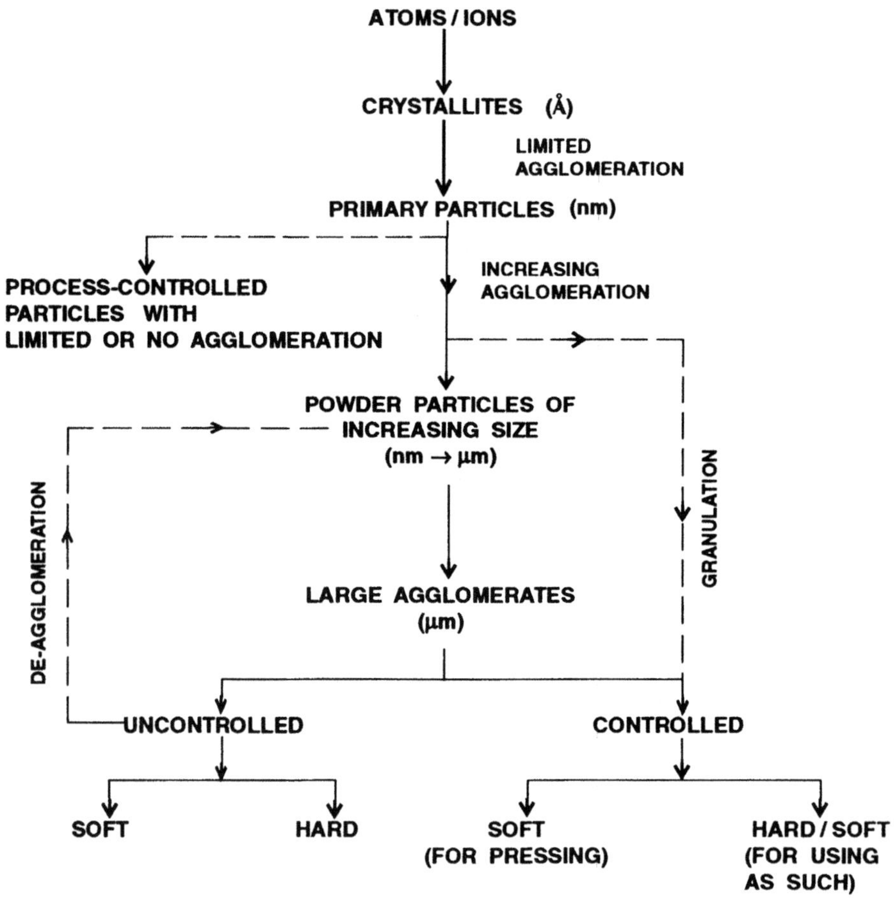

Figure 1.1 Interrelations between crystallites, particles and agglomerates.

which are single crystals made out of a few unit cells. These crystallites can be imagined to be organized among themselves in a co-operative and systematic fashion to form a single crystal primary particle, or simply in a random way to form a polycrystalline one. The primary particles are thus discrete units of single crystals, or mono- or multiphase polycrystals with low or no porosity [1, 2].

An *agglomerate*, i.e. a *secondary* particle has been defined as a small assemblage of primary particles, forming a "network of interconnective pores" [1]. When this assemblage is suspended in a liquid, it is called a *floc* and the phenomenon, *flocculation*. Agglomerates can be uncontrolled in size and shape, i.e. as obtained most frequently during synthesis, or controlled in size (and shape) when the particles are made to go through processes of granulation [1]. Unless the agglomerates are artificially reduced to the practically attainable smallest units (which may go down to the primary particles), the "average agglomerate size" of a freshly synthesized powder is indeed the same as its "average particle size".

Depending upon the number of primary particles taking part in the formation of an agglomerate and the nature of their packing, the size and porosity of the agglomerate change. The strength of an agglomerate depends on the strength of the associated (adhesive) forces, which will be described below.

1.1.2 The Competitive Forces in Agglomeration during Synthesis

Various forces, individually or collectively, can be responsible for the formation of an agglomerate of primary particles. They are [3–5]: electrostatic forces, van der Waals forces, liquid bridges, capillary forces, solid bridges and polymer bridges. Currently some additional forces, called "hydration" or "solvation" forces are also being considered [5–7].

Some or most of these forces are operative during or after the formation of a powder. When agglomerates of particles are formed due to weak attractive forces, they are called "soft" agglomerates which can be broken down to primary particles or at least much smaller agglomerates by light mechanical forces or ultrasonication. When agglomerates are formed through solid bridges between particles, they are called "hard" agglomerates (or *aggregates* when they are very large in size, say, a few mm). Hard agglomerates are specially undesirable in powders for fabrication of ceramics as they do not disintegrate into primary particles by simple means, and need strong mechanical forces for breaking down to smaller agglomerates.

It is well-known that oxide particles dispersed in a liquid, e.g. at early stages of growth in an aqueous solution possess charged surfaces. This happens due to the surface hydroxylation and transformation of the acquired surface hydroxyl groups depending on the solution pH (Fig. 1.2).

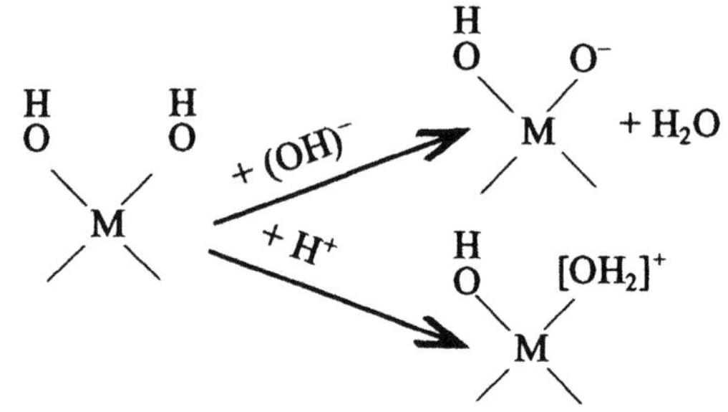

Figure 1.2 Effect of pH on the hydroxylated oxide particle surfaces.

The surface charge attracts counterions from the aqueous solution or the opposite charge of the dipolar molecules (here water) of the mother liquid. This leads to the formation of a layer of progressively changing charge concentration in the solution, with the maximum at the particle-liquid interface. The charge concentration of this layer, thus, decreases with increasing distance from the interface till it equalizes with the concentration in the bulk solution (Fig. 1.3). This layer of changing charge concentration is divided into two parts :

(a) The layer where counterions and oriented dipoles are adsorbed on the particle surface (called the Stern layer). This layer is ready to shear itself away from the bulk liquid and move with the particle which may be, for example, in Brownian motion in the liquid.

(b) The layer next to the Stern layer, called the diffuse layer, in which the (opposite) charge concentration is high, though marked by a sharp gradient.

Layers (a) and (b), taken together, form the so-called "double layer" (Fig 1.3). The thickness of this layer decreases with increase in counterion concentration in the solution. The charge concentration (or potential) which decreases to zero at the outer surface of the double layer is also called the zeta potential; it is conveniently measured from the surface of the Stern layer [8]. The pH at which the zeta potential is zero is called the isoelectric point (IEP); this is conceptually different from the point of zero charge (PZC) indicating the pH at which the particle surface is neutral.

When two particles in a solution come close due to random movements or forced conditions like evaporation of the dispersing liquid, the double layers first overlap. As the double layers of the two particles exhibit concentrations of the same charge, electrostatic repulsion takes place; this hinders a possible contact of the particles and hence, the formation of an agglomerate. Note that the degree of this repulsion is much less than that when two bare surfaces are brought close to each other; the counterions

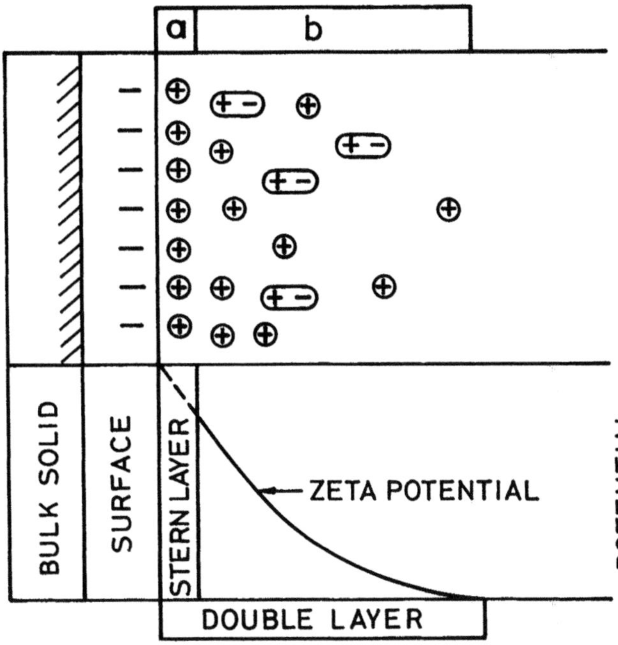

Figure 1.3 The electrical double layer near the charged surface of a particle in a solution.

contribute to this moderating effect. It has been predicted [5], though under a variety of assumptions, that this repulsion bears an approximately exponential relationship with the separation between two (flat) surfaces. When the particles approach each other very slowly, the isothermal work done in bringing them from infinity to the position of closest approach will be the same as the repulsion energy V_R.

The electrostatic force described above is opposite in nature to the van der Waals force, well-known as a small, short-range, attractive force between two isolated molecules which may or may not possess permanent dipoles [9]. The concept has been extended to colloid surfaces for understanding the role of this force in the agglomeration of fine particles. It was proposed by Hamaker [10] that the energy of attraction V_A between two equal spheres of radius a, situated at a distance h (which is the closest approach in vacuum) was

$$V_A = -A\,[(x^2+2x)^{-1} + (x^2+2x+1)^{-1} + 2\ln(x^2+2x)/(x^2+2x+1)]/12 \qquad \ldots (1.1)$$

when $x = h/2a$ and A is the so-called Hamaker constant. The Hamaker constant can be expressed as $A = (\sqrt{A_2} - \sqrt{A_1})^2$ ... (1.2)

where A_1 = Hamaker constant for the liquid, and

A_2 = The same for the particles.

In the above approach followed by Hamaker and others, it was considered that the pairwise interactions between the surface "molecules" are additive. An alternative approach is due to Lifshitz, which does not make the above assumption, but instead depends on the bulk properties of the bodies under consideration. This macroscopic approach of attractive forces has also been applied for colloid systems [11].

For a pair of particles synthesized from and remaining in an aqueous solution (or in other similar polar liquids) the two main factors which decide the nature of interaction between them are therefore the repulsion energy, V_R and the energy of attraction, V_A. The total energy, V_T in this case is simply given by ($V_R + V_T$). The effects and interplays of these two factors have been considered independently by *D*eryaguin and *L*andau in Russia and *V*erwey and *O*verbeek in Holland during the 1940's. The outcome is what is popularly known as the DLVO theory. The discussion below rests on the basic aspects of the DLVO theory.

For a colloid dispersion to be stable for a significantly long time period, the repulsion forces must largely exceed the forces of attraction, so that an energy barrier is set up which would be several orders of magnitude larger than the thermal energy of the colloids. Figure 1.4 shows how the two opposite forces may act in a dispersion of ~ 0.1 – 1.0 μm sized particles in a low-concentration solution of a 1 : 1 electrolyte [12]. It is convenient to remember here that the repulsion energy, V_R acting between two particles decreases in a nearly exponential fashion with the distance between them and the energy of attraction, V_A is approximately inversely related with the square of the distance. The two curves V_A and V_R assume their shapes accordingly.

The sum total V_T, generally dominated by the shape of V_R, exhibits two energy minima in the path to be traversed by two particles approaching each other from a long distance. First comes the *secondary minimum* at which the attractive energy is low, and therefore, the particles form a "non-touching but interactive network" [7], or in other words, a weak agglomerate. When the distance is further reduced, the repulsion energy dominates till a maximum is reached (the energy barrier, which must be crossed by the particles for a strong attachment). With closer approach, the potential energy comes to a very low value called the attractive well or the *primary minimum*, at which a strong attractive action is experienced by the particles. This is the energy regime of formation of strong agglomerates.

With dilute solutions, the double layer becomes the dominating factor and controls the V_T curve. Under these conditions, V_T (max), i.e. the maximum on the V_T curve, which is also the potential energy barrier, remains much larger than the kinetic energy kT of the dispersed particles. When the concentration of ions in the solution is high, the thickness of the double layer diminishes, and the par-

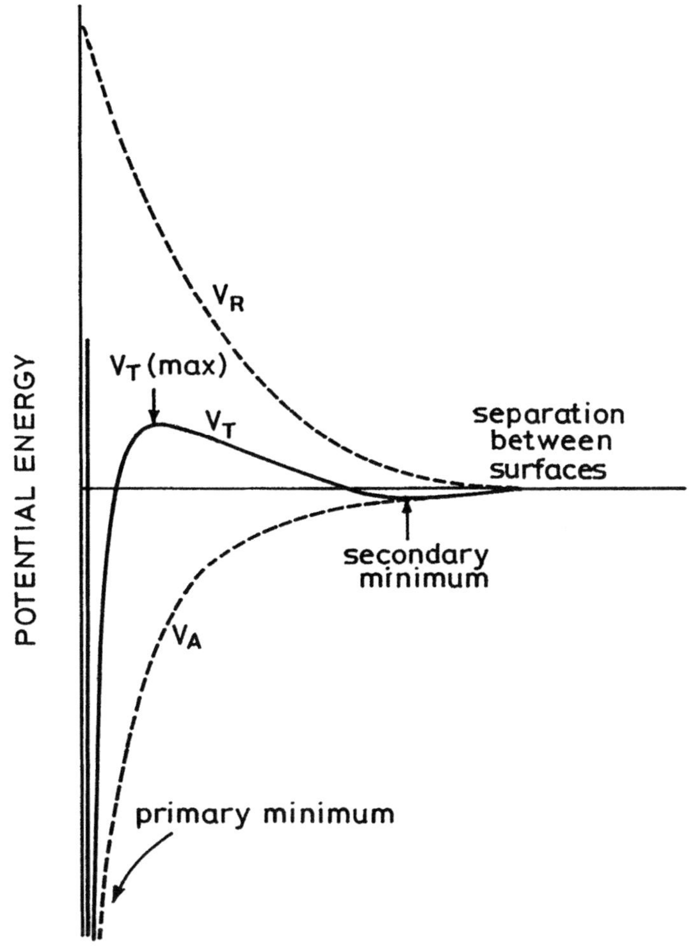

Figure 1.4 The effects of the electrostatic repulsion and van der Waals attraction forces on the potential energy between two particles.

ticles come closer; the result is, of course, the domination of the van der Waals attractive force on the particles, as the kinetic energy $kT >> V_T$ (max). As a result, a strong agglomeration occurs. In addition, the DLVO theory also teaches us that the potential barrier decreases with decreasing particles size, i.e. finer particles get more agglomerated.

Note that when powders (metal hydroxides, hydrated oxides, oxides) are precipitated from homogeneous solutions of metal salts, low solution concentration will lead to the desired low degree of agglomeration, but also a low yield, making the process economically unremunerative. On the other hand, powders with high yield, but also uncontrolled agglomeration (Fig. 1.5) require to be de-agglomerated to

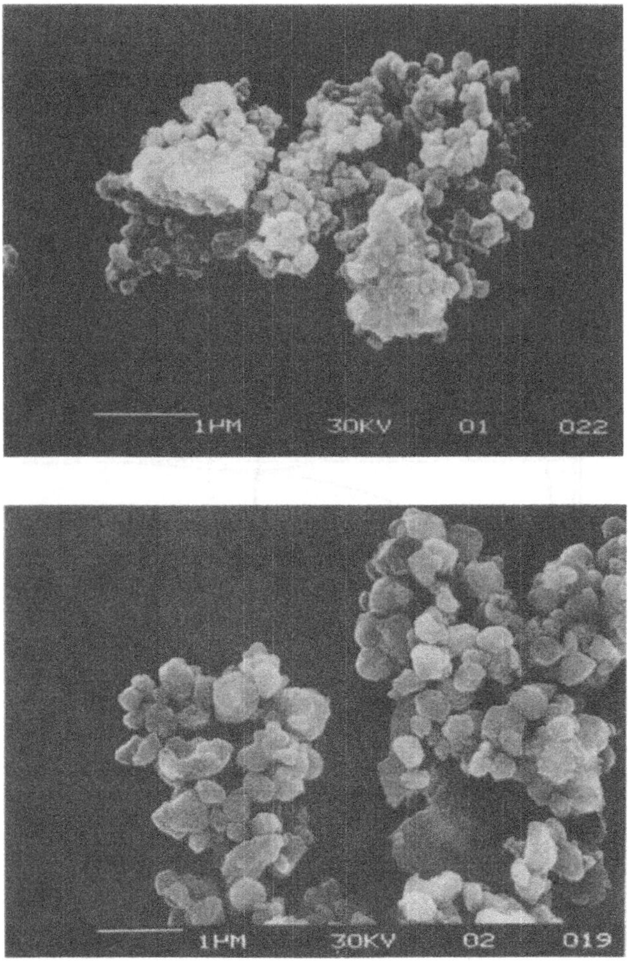

(A)

(B)

Figure 1.5 Agglomerates in submicron zirconia powders obtained from a solution. (A) strong agglomerates; (B) weak agglomerates. Note that the strongly agglomerated particles are smaller than the weakly agglomerated ones.

increase their sinterability. This is often done before calcination at high temperatures to prevent the formation of solid bridges (see below), and hence, hard agglomerates. The other factor discussed above, i.e. the particle size in the precipitate, can be controlled in the solution by various ways, e.g. by selecting the time span of reaction, adjusting the chemistry of the solution etc. [e.g. 13–15] .

Liquid bridges between particles are formed when they undergo drying after synthesis from an aqueous solution (or later, after re-dispersion and removal of the liquid phase). The residual liquid at the contact region of the pair of particles exerts a capillary pressure, leading to an attractive force between the two particles. This needs a little more elaboration. When two particles are in close proximity, the thin space between them acts as a capillary, in which a vapour may condense. When the condensed product, i.e. the corresponding liquid can "wet" the given solid surface [16, 17], a liquid bridge is formed. The critical gap between the two particles, required for the bridge to form, is given by r_K, the Kelvin radius [5] :

$$r_k = \frac{\gamma V}{RT \log(p/p_s)} \qquad \text{... (1.3)}$$

where γ = the surface tension
V = the molar volume
R = the gas constant
p/p_S = the relative vapour pressure of the liquid phase

when $p/p_S < 1$, $r_K < 0$, so a concave meniscus (negative curvature) is expected. This gives rise to a negative Laplace pressure

$$\Delta P = \gamma / r_K \qquad \text{... (1.4)}$$

The attractive force due to aqueous bridges can be of about the same order of magnitude as the van der Waals force (though several times stronger) acting on submicron particles [3].

Solid bridges may be formed in two different ways :

(i) When the liquid bridges formed from salt solutions are dried to the stage of precipitation of the dissolved salt crystals at the neck region between two particles very close to each other.
(ii) When the synthesized particles undergo calcination at elevated temperatures which leads to the transport of material at the neck region between two particles in contact (Fig. 1.6), as during the initial stage of sintering.

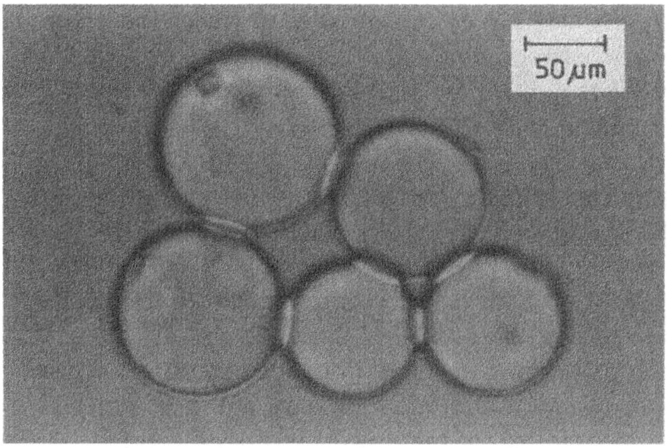

Figure 1.6 Solid bridges between sol-gel derived silica spheres.

Polymers have been used very widely to coat particles in solution by adsorption on the particle surface. A close approach of the polymer-coated particles towards each other causes an overlap of the polymer layers (also called polymer bridge), generating the so-called "steric" forces of repulsion. For creating this, one of the conditions is that the polymer must not be soluble in the dispersing medium (generally non-aqueous solvents are considered here). A surface saturation of particles by polymers which have extended loops is recommended [7] so that a particle-particle linkage does not occur through polymer loops. The use of block co-polymers, with one end in the solvent and the other adsorbed on a particle surface is known to be useful [5]. Thus, when properly adsorbed on the particle surfaces, the polymers give rise to what is known as steric stabilization.

Various operational parameters must be taken into consideration when prevention of agglomeration during solution precipitation or the post-precipitation stage, or de-agglomeration of uncontrolled agglomerates is planned. The importance of such a task in the preparation of powders for ceramic applications cannot be overemphasized, as agglomeration in the starting powder is known to lead to differential densification, inhomogeneous microstructure, low sintered density and poor strength of the ceramics made out of them [e.g. 4, 18, 19]. The important operational parameters are listed in Table 1.1, some or all of which may require consideration for obtaining an agglomerate-free or low-agglomeration powder (which can be converted, if necessary, to controlled agglomerates; Fig. 1.1). The individual points will not be discussed here, as the reader will become familiar with most of these parameters and their importance while going through the subsequent Chapters 3 and 4.

It must be noted that agglomeration is not limited to powders under precipitation from a solution. Some of the forces of attraction described above also act on apparently dry powders, e.g. by condensation of water vapour from the atmosphere in the narrow gap between two particles in close proximity and the formation of a liquid bridge. Thus, particulate products of vapour phase synthesis can also be obtained in agglomerated form, as in the case of SiC [20–22]. Particles under comminution or mixing (specially by wet milling) commonly get agglomerated as they become finer in size. Further, for obtaining a green body for sintering, also out of "dry" synthesized powders, the particles are often dispersed in a liquid (e.g. for colloidal processing). Unless some of the parameters of Table 1.1 are judiciously taken into consideration, agglomerate-free (rare) or weakly agglomerated (common) particles obtained during synthesis now tend to agglomerate in the liquid. The liquid is generally aqueous in nature, distilled or deionized water being the cheapest and safest solvent, unless a very high-purity water is demanded by the end use of the powder, which increases the cost. The above is also true for powder preparation from aqueous solutions.

Table 1.1 Major operational parameters to be considered for obtaining agglomerate-free powders starting from solutions

SOLUTION PRECIPITATION
Choice of starting materials
Proper method and sequence of mixing
Optimum solution concentration
Addition of surfactants in solution (generally emulsions)
Optimum time and temperature of reaction
Optimum pH of the system
Optimum agitation (stirring, ultrasonication)
Controlled rate of particle growth
Controlled level of particle growth
Time of ageing of particles in solution
Controlled agglomeration (spray drying etc.)
DRYING AND CALCINATION OF WET PRECIPITATE
Removal of anions / dissolved salts from particle surface
Replacement of adherent water by organics
Hydrothermal treatment
Freeze drying
Choice of calcination temperature and time
Uniformity of calcination temperature
DE-AGGLOMERATION OF CALCINED PRODUCT
Separation of large agglomerates by sedimentation
Milling
Ultrasonication
Use of dispersant
Use of high forming pressures

When particles are to be dispersed in a solvent, help is often sought of surface active agents (in brief, "surfactants"), named thus because these molecules tend to get adsorbed on surfaces. The most important feature of the surfactants is that they are composed of groups having reversed solubility characteristics : a polar, water-soluble (hydrophilic) ionic group, called the "head", and a non-polar (hydrophobic) hydrocarbon chain, called the "tail" [5, 23]. When the surface of a particle suspended in a liquid is polar, the "head" of a surfactant molecule gets adsorbed onto it due to polar-polar affinity; the reverse is also true. Adsorption of surfactants on particle surfaces, when suitably engineered, can disperse them by altering the chemical nature of the surface and the surface forces.

The family of surfactants is indeed widely extended and offers a large number of options of use. Among the methods of powder preparation, the emulsion process makes a liberal use of different surfactants. Further discussion on these molecules is therefore made in Section 3.3.6.

1.2 APPLICATIONS

1.2.1 Materials for Bulk Sintered Products

Much of the synthetic ceramic powders is used in the fabrication of bulk ceramic products which are, generally speaking, required for advanced or high-performance applications. Two major areas are recognized : (a) electronic ceramics, which include substrates, IC packages, multilayer capacitors, insulators, piezoelectric materials, ferrite magnets and superconductors; and (b) structural ceramics, which include wear parts (mechanical seals, valves, guides etc.), cutting tools, energy applications (heat exchangers etc.), engines (generally ceramic matrixcomposites) and bioceramics. In addition, there are also other important application areas like catalytic converters and ceramic membranes.

It may not be out of place to mention here that all bulk ceramics are not made directly by the compaction and sintering of ceramic powders. A well-known example is the so-called Lanxide process which is based on directed oxidation of a molten metal by a gas [24], leading to the formation of a solid body. Monolithic ceramics can also be prepared directly by the deposition of solid products on a substrate out of reactions of gas mixtures (e.g. chemical vapour deposition). Like other deposition techniques, the thickness can be built up by continuing the deposition process unless the microstructure develops in an unfavourable way (e.g. weak interfaces).

In addition to the above, there are process technologies which use compacts of ceramic precursor(s) for sintering, so that reactions to form the desired ceramic compound and the sintering of the compact can take place simultaneously. The term "reaction sintering" originates from this basic principle of fabrication. Fabrications of monolithic Si_3N_4 from Si powder compacts and N_2 gas, and $BaTiO_3$ from barium carbonate and titania are examples.

In spite of the available alternative options, a major part of the high-performance bulk ceramics are manufactured by the direct processing of fine ceramic powders. Some examples are given in Tables 1.2 and 1.3.

Table 1.2 Some electronic and electrical application areas requiring ceramic powders

Application Area	Ceramic Materials
Capacitors	$BaTiO_3$, other titanates and zirconates, Pb-based "relaxors"
Piezoelectric transducers	$Pb(Zr, Ti)O_3$, $BaTiO_3$
Substrates, packaging	Al_2O_3, AlN, SiC, cordierite
Magnets	Ferrites, high-T_c superconductors
Electro-optics	$(Pb, La) (Zr, Ti)O_3$
Thermistors	$BaTiO_3$
Varistors	ZnO, TiO_2
Solid oxide fuel cells	ZrO_2
Spark plugs / insulators	Al_2O_3

Table 1.3 Some mechanical, structural and other application areas requiring ceramic powders

Application Area	Ceramic Materials
Mechanical seals, valves etc.	ZrO_2, Al_2O_3
Engine components	Si_3N_4, ZrO_2
Cutting tools	Al_2O_3, ZrO_2, TiC, Si_3N_4, Sialon
Bearing materials	Si_3N_4, Sialon, ZrO_2
Nuclear materials	UO_2, PuO_2, ThO_2, SiC, B_4C
Catalytic converters for automobiles	Cordierite
Bioceramics	Al_2O_3, hydroxyapatite, ZrO_2
Sodium vapour lamps	Al_2O_3

1.2.2 Feed Materials for Coatings

Protective ceramic coatings on metals deposited by the thermal spray techniques [25, 26] (plasma spraying being the most important of them when high temperature ceramics are concerned) are gaining importance as advanced materials. Two major characters differentiate the powders for plasma spraying from other kinds of powders of the same substance. One of them is the size range. The temperature in a plasma flame can go up to, say, 20,000°C and the dwell time of a single member of a powder in the flame is less than a millisecond [27]. It is therefore necessary that the powder neither evaporates (too small size and too high temperature) nor leaves the flame in only a partially molten state (too large size and too small dwelling time). For obtaining complete melting, a working size range, different for different substances, but generally within the broad range of 5–100 μm, is recommended by the spray gun manufacturers. This necessitates the manufacture of the powder by controlled agglomeration or by crushing and size classification of agglomerates or fused products.

The other character (though not always observed even in commercial powders, specially when small in size) is a certain degree of flowability of the powder; this ensures a smooth flow through the feed system. The porosity and density of the agglomerated particles also contribute to the performance. Unlike controlled soft agglomerates required for die pressing, those for plasma spraying should not be a loose and low-strength assemblage of particles which may disintegrate into undesirable small-sized units, vaporizing before melting in the flame.

The most important powder candidates for thermally sprayed coatings are listed in Table 1.4.

Table 1.4 Some application areas requiring ceramic powders for the deposition of coatings

Application Area	Ceramic Materials
Wear/erosion resistant coatings on textile machinery, jet engine and automobile parts, pump components etc.	Al_2O_3, Cr_2O_3, TiO_2, Cr_2O_3–TiO_2, Al_2O_3–TiO_2, carbides
Thermal barrier coatings on gas turbine engine parts, diesel engines, aircraft parts	ZrO_2–Y_2O_3, magnesium zirconate
Electrical insulation coatings on high voltage components, computer systems etc.	Al_2O_3, Al_2O_3–TiO_2
Bioceramics (coatings on implants etc.)	Al_2O_3, hydroxyapatite

1.2.3 Pigments, Catalysts, Abrasives and Lubricants

Many high temperature oxides (single or multiple) in the form of fine particles are used as pigments for imparting colour (including white) to paints, glazes etc. [e.g. 28–30]. Submicron particles of crystalline titania provide the best known white pigments because of the high refractive index and non-absorbing optical behaviour (in the visible range), and hence, the highest opacity. It is thus produced in bulk for pigment applications.

The non-white pigments are broadly of two kinds :

(i) Pigments generating (solid) "solution colour", i.e. where a host structure harbours the colouring ion. Typical examples are pigments of yellow/blue colour in which vanadium ions occupy structural sites in zircon or ziconia structure. Development of stains by the incorporation of vanadium or praseodymium in the zircon structure is a long-standing subject of investigation [31–33]. A large number of spinel structures containing transition metal ions (AB_2O_4, where A = Co, Cu, Mn, Ni, Zn etc. and B = Al, Cr, Fe etc.) are used in glaze formulations for imparting various colours [34].

(ii) Pigments constituted of ceramic particles which are coated with colouring matter. Vanadium oxide deposited on tin oxide particles is an example.

The established colour formulations are undergoing changes because of new regulations in the different countries on the use of e.g. chromium and vanadium compounds.

Another industrial area where ceramic powders are used in large quantities is catalyst technology [e.g. 35, 36]. Catalysts, traditionally fine particles of metals and metal oxides, are dispersed uniformly on inert, thermally stable surfaces (catalyst carriers) to obtain high performance. Ceramic powders are required to prepare such catalyst carrier ceramics with high levels of internal surface and porosity for dispersing the catalyst particles by slurry/solution impregnation and other means. The pore diameter may vary widely, i.e. ~ 5 nm – 500 μm according to requirements, and the surface area may be as high as 300 m^2/g.

Alumina in the α- and γ-forms is a common choice as catalyst carrier; γ-Al_2O_3 becomes specially advantageous when it is relatively stable against $\gamma \rightarrow \alpha$ conversion. Alumina-silica mixtures are also widely used; other candidates are titania, zirconia and silicon carbide. Typical catalyst carriers are millimetre-sized beads, hollow cylinders etc. Another area where ceramics have come into use is the conversion of auto-emission products; the ceramic bodies, commonly extrudates of powder slurries, are of course much larger (100 – 150 mm) in size. Cordierite is a well-known candidate [37].

A further application of ceramics in catalysis is in the form of support materials, i.e. spheres or pellets of 2.5 – 50 mm size which support the catalyst bed and control the homogeneity (temperature, reactant flow) of the system. Again, α-Al_2O_3 and Al_2O_3-SiO_2 are in wide use.

Abrasives are generally manufactured in one of the three major forms [38, 39] : (i) coated abrasives (particles coated on paper, polyester etc.), (ii) bonded abrasives (e.g. grinding wheels) and (iii) abrasive grains, generally of relatively large size.

Early abrasive grains or particles were natural minerals like flint or garnet, as also man-made fused alumina or silicon carbide grains; the last two are the mainstays of the abrasive industry [38, 39]. Fused alumina-zirconia grains became commercially available in the early sixties. The sol-gel technique was first employed during the early to mid-eighties to produce pure and doped alumina [40], as also Mg-spinel [41].

Intensive proprietary work has been carried out on sol-gel derived, fine-particle alumina abrasives, as also alumina with zirconia and some rare earth oxides [40, 42]. Aluminium oxynitride has also been suggested as an abrasive material [43]. Among the other relatively new abrasives, alumina-titanium nitride is known for coated / bonded abrasive applications. Cubic boron nitride is one of the hardest substances and is therefore a strong candidate for abrasive applications. However, the powder is generally produced from the much softer hexagonal form at high temperatures and pressures (1200°–1700°C, ~ 6.5 GPa), which may restrict its use.

Lubricants required for room temperature or slightly elevated temperatures are generally liquids. For high temperature applications, relatively soft ceramic powders are required. The other property of such powders is a flaky or layered morphology which comes from the crystal structure. Natural graphite flakes are the commonest example [44]. Other solid lubricants are molybdenum disulphide and hexagonal boron nitride. Graphite and MoS_2 are used at temperatures not exceeding about 400°C. Hexagonal BN, which is structurally akin to graphite (hence the name "white graphite"), can be used up to 800°C [45]. The powder can be used as such, i.e. dry, but also as a suspension in an organic liquid or as a spray [46].

Examples of the applications for the above and other powders are summarized in Table 1.5.

Table 1.5 Some conventional areas requiring ceramic powders

Application Area	Ceramic Materials
Pigments	TiO_2, ZnO, $ZrSiO_4$, ZrO_2
Catalysis	Al_2O_3, Al_2O_3–SiO_2, TiO_2
Abrasives	Al_2O_3, Al_2O_3–ZrO_2, SiC, B_4C
Polishing media	Al_2O_3, CeO_2
Lubricants	Graphite flakes, hexagonal BN
Porcelain insulators, enamels etc.	Feldspar, kaolin, ball clay, SiO_2, Al_2O_3

1.2.4 Bioceramic Materials

The most widely used bioceramic materials are alumina and the calcium phosphates (specially hydroxyapatite), as also partially stabilized zirconia and various bioglasses [47]. The applications depend on the function the material (in the required form) is expected to perform. Thus, alumina and zirconia are *bioinert* materials, one of the major applications of which is as implants. Hydroxyapatite can attach itself to the bone directly by chemical bonding, and is *bioactive*. Some members of the calcium phosphate group, e.g. β-tricalcium phosphate are slowly replaced by bone, and are *resorbable* ceramics [47].

Compositional and phase purity is an important consideration for bioceramic materials. One significant example is alumina ceramics for implant applications. The initial implants (1970's) used alumina of about 99.1 wt% purity, but in about 15–20 years the desired purity level has gone up to 99.8 wt% and beyond [48], while impurities like SiO_2 and MgO have gone down to a total of less than 0.1 wt%. Partially stabilized zirconia requires at least 97 wt% ZrO_2 [47], so the dopant content must be very low.

Commercial powders of hydroxyapatite are often not phase-pure, i.e. single phase [49]. For the desired powders, the requirements are, for example, to avoid pyrophosphates which inhibit the formation of hydroxyapatite and amorphous and hydrated impurities which may reduce the strength of the ceramic.

2

WHY SYNTHETIC CERAMIC POWDERS ?

2.0 INTRODUCTION

At the beginning of this Chapter we propose to remind the readers that

(i) For many ceramic compounds which have found almost irreplaceable position in industrial applications or are future candidates for the same, there are no natural equivalents. Elegant processes with a high degree of control have been developed for the synthesis of almost all such kinds of crystalline compounds in the form of fine particles to take care of this inadequacy, as will be revealed in Chapter 4. Although synthetic processes are sometimes directly dependent on natural raw materials, most synthetic powders are prepared from relatively pure chemicals (including acids, bases and solvents); the degree of purity is dependent on the envisaged use.

(ii) Natural minerals are generally impure, and attempts to increase the purity level to match specific requirements often prove to be unsuccessful or economically unattractive. On the other hand, synthetic powders can be extremely pure and intentional addition of impurities in them can be precisely controlled in many processes of synthesis.

(iii) In spite of vast improvements in the milling techniques, powders obtained by the milling of natural minerals still fall short of expectations regarding fine size, size distribution and shape. Synthetic processes can now produce particles with controlled shape and size, and narrow size distribution [14, 50–52].

Critical failure analyses of advanced ceramic products have shwon that microstructural flaws in ceramics are often caused by low-quality and non-uniform starting powders [18]. This can be avoided by using only synthetic procedures of powder preparation with a strict process control.

2.1 INADEQUACIES OF NATURAL POWDERS

2.1.1 Compositional Variations

Natural raw materials for ceramics were formed in the old geological times through millions of years by the crystallization of generally multicomponent dry or aqueous melts into rocks, and the subsequent/concurrent action of hot solutions and vapours, or simply weathering action of the natural agencies on the formed rocks. All these processes involved the interaction of a large number of ionic constituents. As a result, natural minerals are by and large impure in the as-received form, which is reflected in their chemical analyses. The impurities may be structural (substitutional or interstitial solid solutions etc.) or non-structural (inclusions within mineral grains, incrustations on the external surfaces or within the crevices etc.). While the non-structural impurities can be reduced or removed by conventional means, e.g. acid washing, reduction or removal of structural impurities is often difficult. The cost of purification, of course, depends on the quantum and type of the impurities. Some examples showing the levels of impurities in natural raw materials for ceramics are discussed below.

Bauxite is the most important and abundant raw material for alumina powders. It is a product of chemical weathering of aluminous minerals in different rocks and the accumulation of the aluminium compounds thus formed. The main constituents are gibbsite, $Al(OH)_3$, diaspore, $Al(OH)_3$ and boehmite, $AlO(OH)$; the relative predominance changes with the origin. Apart from this, bauxite contains other mineral constituents and impurities. A summary of the ranges of contents of the principal elements in some important Indian bauxite ores [53], given in Table 2.1, indicates the total impurity that would require removal before a pure gibbsite is obtained (Section 4.1.1).

Table 2.1 Ranges of the major constituents of some Indian bauxites [53]

Oxide	Range (wt%)
Al_2O_3	48.57 – 60.75
SiO_2	1.28 – 7.27
Fe_2O_3	1.85 – 13.60
TiO_2	0.30 – 10.35
Volatiles	23.80 – 32.30

Zircon, $ZrSiO_4$ is the most important raw material for the production of zirconia powders [54]. It is also a direct input in some refractories, and is currently being considered for advanced ceramic applications (Section 2.2.3). Available as individual sand grains (Fig. 2.1) in coastal deposits of some countries [55], it is relatively free from associated mineral impurities (as aganist ores obtained from

Figure 2.1 Optical micrograph of individual grains from an Indian zircon sand deposit.

multi-mineral rocks), but can contain impurities in the form of incrustations, inclusions and substituted cations. The ranges of contents of the principal elements present in zircon ores from different parts of the world are shown in Table 2.2. The variations are sufficiently high to be unacceptable for direct applications as proposed by, e.g. Shi et al. [56].

Table 2.2 Ranges of the major and minor constituents of natural zircon from some different sources [54–56]

Oxide	Range (wt%)
ZrO_2 (+ HfO_2)	58.06 – 65.82
SiO_2	31.70 – 33.42
Fe_2O_3	0.06 – 0.50
MgO	0.02 – 0.11
TiO_2	0.04 – 0.44

Cordierite, $Mg_2Al_4Si_5O_{18}$ is another mineral which can be directly used for advanced ceramic applications, but for the relative rarity of the natural occurrences of this mineral and its low purity level. The main impurities in natural cordierite are the iron oxides, evident from compilations of chemical analyses of the mineral both from widely distributed [57] and restricted [58] localities (Table 2.3).

Table 2.3 Ranges of the major constituents of natural cordierite from different sources

Oxide	Range (wt%)	
	[57]	[58]
MgO	1.48 – 12.16	6.6 – 12.3
Al_2O_3	28.72 – 35.25	32.7 – 35.1
SiO_2	40.92 – 54.63	48.8 – 50.2
(FeO + Fe_2O_3)	2.26 – 16.93	2.2 – 11.3

2.1.2 Particle Size Distribution and Shape

The size distribution in natural powders is one feature which can have similarities with the same of many of their synthetic counterparts. The reason behind this is that natural powders are nearly always products of milling (Section 3.1) and classification, which is also the route required to be generally followed for industrial synthetic powders. This is especially true for cases where fine synthetic powders are manufactured via extensive milling operations of large chunks or simply agglomerates of the synthesized compound, e.g. SiC by the Acheson process (Section 4.2).

Synthetic powders, however, enjoy several advantages over the natural ones :

(i) The nucleation and growth phenomena of synthetic powders can be suitably monitored to optimize the generation of a desired size range, down to unagglomerated nanoparticles [15].

(ii) Options are also available for narrowing down the size distribution of various synthetic powders without milling and classification. This includes the posibilities of obtaining nearly monosized powders [e.g. 59, 60].

(iii) Several solution and vapour phase techniques can control the synthetic particle shape [59–62]; the shapes of natural particles are generally angular, which comes from the different milling operations.

2.2 THE NATURE OF CERAMIC POWDERS : MODERN DEMANDS AND DEVELOPMENTS

2.2.1 Powder Size and Size Distribution

The strength of a green body (σ) composed of packed powders has been shown to be inversely proportional to the particle radius (r), as evident from a simplified form of the Rumpf equation [63] :

$$\sigma = \frac{1.1\phi}{1-\phi} \cdot \frac{A}{12rl^2} \qquad \text{... (2.1)}$$

where ϕ = solid volume fraction in the green body
l = distance between the particles (idealized as spheres)
and A is a constant.

All proposed mechanisms of sintering and densification of ceramic powder compacts agree that the particle size is one of the most important parameters in the rate of progress of these processes. Early studies showed that the rate of densification in a simple viscous system is inversely proportional to the particle size; for volume diffusion it is inversely proportional to the 1.5 power of the size [64].

That small particles require less time to sinter is exemplified [65] by simple scaling relationships proposed by Herring [66], well-known as Herring's scaling laws :

$$\frac{t_1}{t_2} = \left(\frac{r_1}{r_2}\right)^n \text{ at constant temperature,} \qquad \text{... (2.2)}$$

where r_1 and r_2 are the radii of two different spherical particles arranged in identical arrays (e.g. hexagonal/cubic close packing) and t_1 and t_2 are the corresponding times required for the two powders to sinter (presumably to the same or similar degrees of densification). The value of n can vary according to the mechanism of sintering. Thus, for lattice and grain boundary diffusion, the values of n are 3 and 4 respectively. Yan [65] pointed out that a decrease in the particle size from 1μm to 0.01 μm would lead to a decrease in the sintering time by a factor of 10^6 to 10^8. Figure 2.2 shows how the relative density of a sintered body can be influenced by

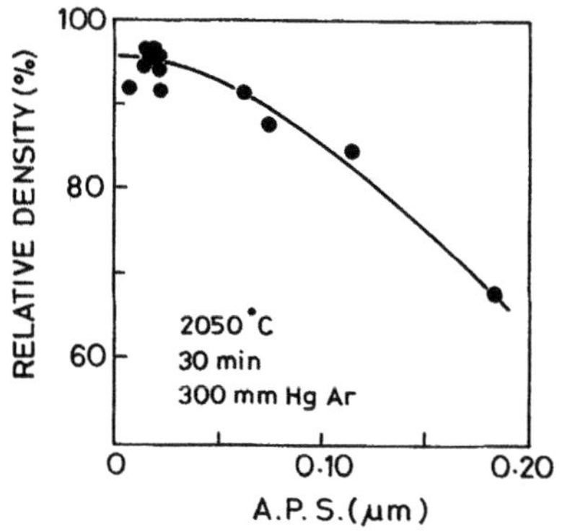

Figure 2.2 Relative density of a sintered body of SiC as a function of the particle size [67]. (Reproduced with permission from the Indian Ceramic Society, Calcutta, India)

the particle size of the starting powder [67]. In case of a reaction among powder particles, smaller particle size leads to higher rates of compound formation, though not as significantly as in the case of densification.

The above observations make it clear why an well-accepted modern demand is for unagglomerated "ideal" powders with an average size within the range 0.1–1.0 µm [e.g. 18]. The present status of powder preparation, in view of the above demand, can be summarized in the following way :

(a) Relatively new processes for the synthesis of micron- or submicron-sized powders (generally from solutions and vapours, see Chapter 4), in different scales of operation with different possibilities of upscaling, have become available through the last 10–15 years. Some of these processes claim facile synthesis of nanoparticles, while others can produce powders > 1 µm.

(b) The modern processes of fine powder preparation relevant to this discussion can be broadly divided into three groups :

- (i) those yielding particles distributed in the size range of around 0.1–1.0 µm, with an average size somewhere in between, i.e. matching the "ideal" character.
- (ii) those yielding particles of very narrow size distribution (may or may not be "monosized") with an average lying within the suggested range.
- (iii) those yielding particles that are entirely finer than the lower limit of 0.1 µm.

In many of these processes enough controlling parameters are available for altering the size range according to necessity [e.g. 13]. Further, note that

- (i) The products of some of these processes are known to be highly sinteractive particles which densify to ~ 99% of the theoretical value without sintering aids and at relatively low temperatures [e.g. 50].
- (ii) In spite of the successful developments in the control of the particle size, the control of agglomeration (during or after synthesis) still remains a key issue. Unagglomerated fine particles obtained through elaborate processes (Table 1.1), adding to the cost of production, can become at least softly agglomerated due to wrong storage and handling (especially in a humid atmosphere).

Thus, there still seems to be some scope for the development of facile and economically acceptable methods of controlling the agglomeration of the fine particles being synthesized by the current techniques for dry pressing. Colloidal processing provides a potential alternative to the dry pressing of agglomerated powders [7, 19], though the initial removal of oversized material by simple sedimentation seems to be a slow process unless done via centrifugation. In both cases, it is a cost-escalating step. Kendall [68] visualized the commercial availability of fine ceramic particles in the form of concentrated dispersion, ready for processing into ceramic bodies. For dry pressing, it is apparently useful to convert starting solutions or dispersions of uncalcined fine particles into spherical soft agglomerates of optimum size to avoid uncontrolled hard agglomerate formation and fulfil the requirement of flowability for die-filling (see Chapter 3). However, it is now known that such agglomerates generate flaws during pressing which are unacceptable for modern ceramics [7].

2.2.2 Particle Shape

The problems associated with uncontrolled powder agglomerates during sintering have been discussed by various authors [4, 7]. Another factor in powders which is now considered important for obtaining ceramics with homogeneous microstructure is the shape of the individual particles. This understanding has developed basically along with the necessity of using synthetic powders of controlled size rather than crushed and milled, irregular and coarse particles. The currently accepted concept [18, 65] is that for achieving enhanced and uniform densification, the "ideal" particles should have an equiaxed shape, i.e. no dimensional preferences.

The simplest equiaxed shape is that of a sphere. This, and the impetus from early colloid research [69] demonstrating that spherical, submicron, monosized and unagglomerated particles can be easily prepared from alkoxide solutions (Fig. 2.3), have led to the preparation of spherical ceramic particles from alkoxide and modified

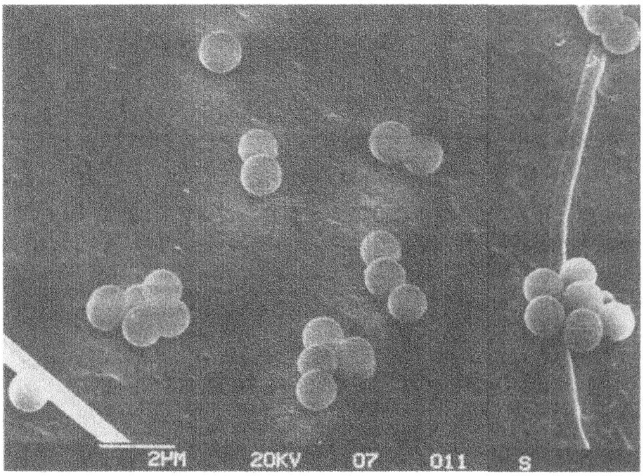

Figure 2.3 Scanning electron micrograph of spherical silica particles obtained from an alkoxide solution [70].

alkoxide solutions [e.g. 14, 50, 59, 60, 70]. In fact, there are now a large number of new synthetic routes available which can generate individual or agglomerated spherical particles from non-alkoxide chemicals through solution or vapour reactions [e.g. 71, 72 for TiO_2]; the details of synthesis are available in Chapter 4. It should be noted at the same time that these particles are generally far from monosized.

It has been argued that submicron, monosized spherical particles, when packed in regular arrays and sintered, should give rise to ceramics of ideal microstructure with similarly sized grains and pores, or grains without porosity. Unfortunately this visually attractive model is yet to be fully realized in practice. Thus, it has been concluded by Chappel et al. [73] that assemblages of monosized (presumably spherical) particles can exhibit the presence of defects like dislocations, domain boundaries and vacancies normally not observed in packings of irregular powders. There are, however, signs [74–76] that improvements in powder processing technology may lead to important advantages in the densification of such special powders. In addition, there are also strong evidences that well-packed equiaxed particles assist in better densification [65] and that particles with jagged boundaries lead to problems in densification, especially through liquid phase sintering [77, 78].

When conditions favourable for crystallization prevail during synthesis, particles are obtained as submicron- or micron-sized individual crystals (as in hydrothermal synthesis, see Chapter 4). Under the conditions of synthesis, or post-synthesis processing like liquid phase sintering, particles may crystallize with distinct morphologies which may or may not represent a particular polymorph. The frequently obtained needle-like morphology of β-Si_3N_4 particles as compared to nominally equiaxed particles of α-Si_3N_4 is an example. The investigations of Matijevic and co-workers show the wide variations of particle morphology that can be produced by controlling the parameters of synthesis [e.g. 79].

Wötting and Ziegler [80] reported that with a low additive content (5 wt% Y_2O_3 + 1.3 wt% Al_2O_3), equiaxed particles of Si_3N_4 showed better sinterability than partly rod-like particles in a liquid phase sintering process. This is, however, not to indicate that rod-like or needle-like β-Si_3N_4 particles have been proved to be an undesirable powder precursor for Si_3N_4 ceramics. One of the current concepts is to take advantage of this elongated shape of variable aspect ratio as a function of the β-content [81] in obtaining in situ fibre reinforcement [82]. A desirable condition, thus, is a starting powder with a high α-Si_3N_4 content (equiaxed) which would help in the interlocking of rod-like β-Si_3N_4 grains in the ceramic [80]. On the other hand, very high (> 95 vol%) β-content can lead to a steric hindrance against the growth of the rod-like morphology, resulting thereby in a fine-grained, homogeneous microstructure composed of equiaxed grains [82].

Another ceramic powder which is obtained as particles of distinctly different shapes as a function of process parameters is ZnO. While equiaxed particles are easy to obtain, chemical conditions are known which can generate particles of needle-, rod- or prism-like morphology [61, 83–85]. Equiaxed particles remain the general choice, but prismatic particles have been recently shown [85] to produce a sintered microstructure suitable for sensor applications; proprietary processes have been developed for the production of acicular particles [86].

2.2.3 Compositional Purity and Homogeneity

The requirements of control over the chemical purity of modern ceramic powders, especially those for use in electronic applications, are quite stringent. This is due to the fact that (i) even minor quantities ($<<1\%$) of certain impurities can exert significant effects on the properties of the corresponding ceramics, and (ii) specific minor impurities are required to be doped (introduced during the synthesis or added later) in essentially pure compounds for obtaining the best results, e.g. doping of Bi, Sb etc. in zinc oxide ceramics for varistor applications [87].

The impurity content of alumina powder provides a classical example of the efforts gone and still going into increasing the purity of ceramic powders following the modern demands. While alumina for the refractories, insulators or glass can afford to contain > 4 wt% of Na_2O, this value comes down by an order of magnitude for structural ceramics [88]. Modern electronic ceramics and tubes for sodium vapour lamps require alumina particles of ≥ 99.99% purity. Such particles are allowed to have only less than 20 ppm alkali and alkaline earth cations, as also within 10 ppm of any individual transition metal ion. A further demand is the elimination of sulphate and chloride impurities [89].

The demand for an increase in the purity of alumina for bioceramic applications has already been mentioned (Section 1.2.4).

For catering to this necessity, processes have been developed for obtaining high-purity powders from the conventional Bayer aluminas, as also other sources [e.g. 89, 90]. High-purity aluminas (99.99% or more) are now under routine production [89–91]. More detailed indications of the recent efforts will be evident from the discussions in Section 4.1.1.

Aluminium nitride, AlN is a modern ceramic material considered to be an excellent candidate for the following advanced applications [92] :

(i) Substrates for hybrid and power circuits, as also multichip modules.
(ii) Packaging materials for integrated circuits.
(iii) High heat conductive plastic package applications, i.e. filler in epoxy moulding compounds for increasing their thermal conductivity.

For the best performance in the first two applications, the powder must be essentially free from metallic impurities (e.g. < 35 ppm Fe and < 100 ppm Si); the residual carbon content must go below 0.1 wt% and the residual oxygen, below 1.2 wt%. The filler requirements are not so stringent regarding the metallic impurities, but insist on a low oxygen content (to avoid any deterioration in the thermal conductivity). As a result, the demands on the purity of the starting materials and a control of the reaction process involved in the production of AlN powders are very high. Several commercial and research powders [e.g. 91, 93] match the requirements almost completely. The major techniques include direct nitridation of Al powder and nitridation with Al/AlN feeds (see Chapter 4).

As in the case of AlN, the requirements of purity are also known for Si_3N_4 particles, e.g. low cationic impurities (< 2 wt%, but preferably < 0.05 wt%) and low oxygen (< 2 wt% but preferably < 1.0 wt%), and a low carbon content (< 0.1wt%) [94, 95].

Further, halogen impurities (F, Cl) coming from the raw materials (e.g. $SiCl_4$), subsequent chemical treatment or fluxing agents [96–98] have been found to be responsible for the degradation of high temperature mechanical properties of the ceramics by decreasing the viscosity of the sintering liquid [98], and hence, must be minimized. Some commercial powders are known to contain more than 12 wt% Cl [98]. Attempts have therefore been made to remove unwanted chlorine from Si_3N_4 powders [e.g. 99].

The issue of oxygen content, especially the surface oxygen content, has been thoroughly examined by surface modification of α-Si_3N_4 powders and their conversion to bulk ceramics [100]. It has been found that increasing surface oxygen in fact leads to increasing modulus of rupture and flexural strength of the ceramics, especially in presence of additives.

An interesting example of a new demand on the purity of synthetic powders is that in respect of zircon, $ZrSiO_4$ which for many years has been used in pigments and refractories where small impurities have been considered unimportant, or sometimes even necessary. Current research [101, 102] is looking at zircon as a potential candidate for high temperature structural ceramic applications. The interest comes from its low coefficient of heat conductivity and thermal expansion, as also high temperature mechanical strength and thermal shock resistance. Though available as beach sand in large renewable deposits in several countries, natural zircon is generally impure (Section 2.1.1). The demand on the purity of zircon for utilizing its potential in full has led to the re-examination of the synthetic route (see Chapter 4).

Shi et al. [56], who prepared zircon powder by a solution/sol route, achieved a much higher purity than that of its natural counterpart. Table 2.4 provides a comparison; according to these authors, a part of the impurities in the synthetic powder came from the natural zircon particles (optimally 7 mol %) added as seed crystals.

Table 2.4 Chemical compositions of natural and synthetic zircon [56]

Oxide	Natural Zircon (wt%)	Synthetic Zircon (wt%)
ZrO_2	62.28	66.90
HfO_2	1.26	1.35
SiO_2	31.70	31.21
Al_2O_3	3.17	0.15
Fe_2O_3	0.14	0.021
TiO_2	0.44	—
CaO	0.27	0.006
MgO	0.081	0.005
Y_2O_3	0.11	—

In addition to the above, there are many instances in the literature where synthetic powders of high purity have been obtained as a matter of routine [62, 89, 103]. The impetus, of course, comes from the envisaged modern applications.

Along with increased demands on the purity of ceramic powders, the compositional homogeneity of individual particles is also being looked into. While the overall purity of the research and commercial powders (with or without dopants) can be adequately high, a highly uniform distribution of the constituents in individual particles seems yet to be achieved. Some examples will explain the position.

Uchiyama et al. [104], who reported EDX analysis of alkoxide-derived Y_2O_3 (nominally 3 mol%)–ZrO_2 spherical powders, showed that the Y_2O_3 content in particles ranging from ~ 0.45 to 0.8 μm was 3.4 ± 0.3 mol %. A similar work by Ishizawa et al. [105] on spray-pyrolyzed Y_2O_3–ZrO_2 powder showed that the Y_2O_3-contents remained more or less constant at ~ 2.3 mol % for particles 0.20-0.35 μm in size, but deviated in case of smaller particles. EDX analysis of (Pb, La) (Zr, Ti) O_3 powders obtained by spray pyrolysis [106] demonstrated that the contents of the different oxides had a narrower distribution in the individual particles than in particles prepared by solid-solid reactions.

2.2.4 Surface Modifications

The role of the particle surface in controlling the agglomeration phenomena has been discussed in Chapter 1. The understanding that the repulsive effect coming from the double layer overlap is no more applicable at the isoelectric point (IEP), so that the particles would flocculate and precipitate under such conditions, is well-known. The preparation of a relatively stable dispersion by adjusting the pH of the system reasonably away from the IEP of the particles in question is also in practice for many years.

The situation takes a different turn when a two- or multiphase particle population is to be processed (as in particulate composites), because particles of the different substances have different IEP values (see Table 2.5 for some examples). Simultaneous flocculation is effected by depositing optimally thick coatings of one sub-

Table 2.5 Approximate values of the isoelectric point (IEP) of some ceramic substances

Composition	IEP (at pH)
Al_2O_3	8 – 9
MgO	12
Mullite ($3Al_2O_3.2SiO_2$)	6 – 8
SiO_2	2
ZnO	9
ZrO_2	4 – 6

stance on particle surfaces of another substance, e.g. TiO_2 on Al_2O_3 particles for effectively obtaining the same IEP values for all the particles in suspension [107]. The green packing thus obtained is supposed to be more homogeneous than flocculated assemblages of particles of two different substances.

Non-oxide powders often require special surface modification because of the proneness of their surface atoms to react with the environment. For example, aluminium nitride is well-known for its very facile hydrolyzability, preventing the suspension of AlN particles in water. The following reactions have been proposed [108–110].

$$AlN + 2H_2O \rightarrow AlOOH\ (amorph) + NH_3 \quad \text{... (2.3)}$$

$$NH_3 + H_2O \leftrightharpoons NH_4^+ + OH^- \quad \text{... (2.4)}$$

$$AlOOH\ (amorph) + H_2O \rightarrow Al(OH)_3 \quad \text{... (2.5)}$$

$$AlN + 4H_2O \rightarrow Al(OH)_3 + NH_4OH \quad \text{... (2.6)}$$

Thus, Bowen et al. [110] proposed the formation of a surface layer of AlOOH, gradually changing to $Al(OH)_3$. They further showed from XPS studies that a commercial AlN powder (average size 1.2 μm, specific surface area 6.2 m^2/g, containing 2.2 wt% O_2), when suspended under stirring in ten times its weight of water, lost almost all of its surface nitrogen atoms after 24 h. Reetz et al. [109], who worked with an AlN powder of size 7.0 μm, specific surface area 0.7 m^2/g and containing 0.96 wt% O_2, showed that its oxygen content increased to ~ 4.5% at pH ~ 11.5, but jumped to ~ 60% when the temperature was raised to 50°C at the same pH.

Considering this problem, and the demand for surface modifications of AlN powder, several manufacturers now produce coated varieties of AlN. The detailed study of Metselaar et al. [108] has shown two of these commercial powders to be coated with silicon-organic carbon and phosphate groups, the latter showing good stability in water. In another recent investigation, Egashira et al. [111] used a commercial AlN powder with an average size of 0.5 μm and specific surface area 3.7 m^2/g, partially dehydroxylated at 150°C. The surface was found to be favourable for the adsorption of carboxylic acid molecules with carbon atoms varying from 8 to 22. The surface reaction proposed was

$$= Al - OH + HOOC\text{–}R \rightarrow = Al - O - \overset{\overset{\displaystyle O}{||}}{C} - R + H_2O \qquad \text{... (2.7)}$$

It was shown that (i) the chemical stability against water increased with increase in the number of carbon atoms in the acid, and (ii) the carboxylic acid coated AlN powders could be dispersed in water. The dramatic effect of particle surface coating by stearic acid [112] on the AlN → $Al(OH)_3$ reaction is clear from Fig. 2.4 (i) and (ii) [112].

The simplest route of incorporation of an additive to a powder before compaction and sintering is through mechanical mixing. In such procedures, the degree of dispersion of the particles of the additive in the batch of the host particles is restricted by, among others, the level of difference in the particle size distribution of the two [113], which can often be quite considerable. Compacts of such inhomogeneous mixtures, on sintering, lead to relatively non-uniform microstructure. When the additives are meant for liquid formation required by the pressureless sintering technique, inhomogeneous distribution of the liquid can cause significant deterioration of the mechanical properites. Co-precipitated powders suffer from contamination by counterions (nitrate, halide etc.), retention of which is also highly detrimental to the sintered properties.

Attention is therefore being paid currently to the possibility of homogeneous distribution of the additive on a "nanoscale" [114] on the surface of each particle as a

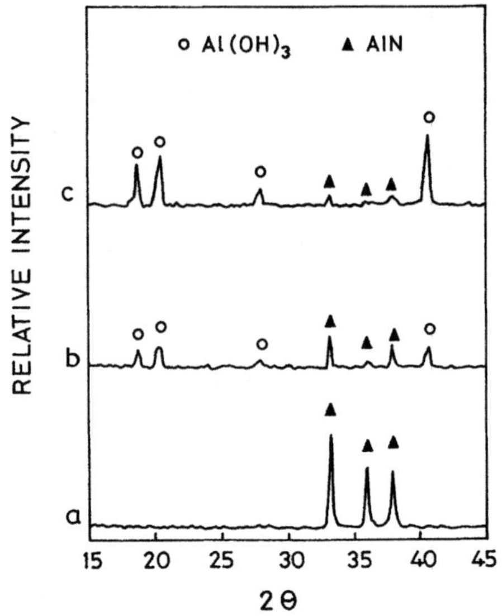

(i) X-ray powder pattern of uncoated AlN powders after
(a) 0-, (b) 24- and (c) 96 h water treatment [112].

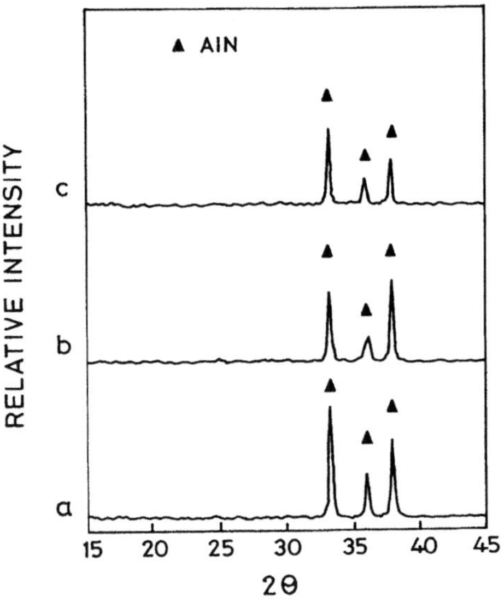

(ii) X-ray powder pattern of stearic acid-coated AlN powders after
(a) 0-, (b) 24- and (c) 96 h water treatment [112].

Figure 2.4 Effect of stearic acid coating on AlN powder.
(Reproduced with permission from the Deutsche Keramische Gesellschaft, Köln, Germany)

thin layer before green body formation and sintering. Some experimental details in case of Si_3N_4 powders (which are also generally valid for other powders) are presented in Table 2.6 to show the current course of developments in this area.

Table 2.6 Examples of experimental parameters used in coating Si_3N_4 powders

Nominal nature of coating	Source chemical for coating	Solvent	Precipitating agent	Reference
Al_2O_3	Isopropoxide	Isopropanol	Water	[115]
"	Nitrate	Water	Urea	[114]
" (AlOOH)	s-Butoxide	Water	Water	[116]
"	Isopropoxide	n-Hexane	Water	[117]
"Al_2O_3"	Nitrate	Water	Urea	[118]
"	s-Butoxide	Hexane	Water	[118]
MgO	Methoxide	Methanol		[113]
"	Nitrate	Isopropanol		[119]
MgO, Al_2O_3	Double butoxide	Isobutanol		[120]
Y_2O_3, Al_2O_3	Nitrates	Water	Ammonia	[121]
Y_2O_3, CeO_2	Isopropoxide	Benzene	Water	[122]
Y_2O_3, MgO	Nitrate	Water	Tetraethyl ammonium hydroxide	[123]

The results of these experiments show that broadly, a nanosized coverage of particle surfaces has been achieved in many cases, but the optimum conditions of uniform deposition are yet to be standardized. The coatings have been described by various workers as discontinuous, porous or wispy. In spite of this, such surface modifications have been shown to improve the sintering behaviour of Si_3N_4 powder [123]. It has been pointed out by Schmidt et al. [113] that for obtaining a homogeneous and thin "oxide" coating on Si_3N_4 particles, it is also necessary to control the surface hydroxyl content : a relatively large specific surface area and surface OH content was considered to be an advantage. One of the simple routes of manipulating the silanol content of Si_3N_4 is ageing of the particles in water [124].

3

TECHNIQUES OF POWDER PREPARATION

3.0 INTRODUCTION

Apart from the classical methods of powder preparation — some of them fifty to one hundred years old — there are now practically innumerable new methods available which lead to high-quality (with respect to size, shape and purity) powders of oxide and non-oxide ceramic materials. Many of these new methods are, again, variants of and/or improvements over the well-known procedures used by the industry. The other point that merits mention is that notwithstanding the claimed high quality of the powders, many of the new and elegant methods deal with only gram-quantities of material and (especially in case of vapour-phase reactions) have a very slow rate of powder generation. This indicates that a success in the upscaling of many such otherwise attractive processes while maintaining the product quality is not guaranteed at the present stage.

Considering the above points, this chapter will introduce in general the most well-known processes of powder preparation, with relatively few mentions about less known, new processes as and when felt necessary.

3.1 TECHNIQUES OF COMMINUTION

Comminution can be described as the technique for reducing large-sized solids into smaller ones (grains, particles) by mechanical forces, e.g. by crushing, grinding and milling. In the discussions that follow, the crushing operation will not be discussed as they produce only ~ 1 mm (or fractions thereof) sized grains which are not considered here. Also, the terms "grinding" and "milling" will be used somewhat synonymously as a range of processes producing micron- to sub-micron-sized powders [125].

Grinding of crushed and classified feedstock of minerals or hard synthetic aggregates into relatively fine (as above) powders proceeds through several basic (but not always exclusive of one another) mechanisms : compression, mechanical impact and particle attrition (wear) at high velocities. In case of compression, the relatively coarse feed material undergoes stress between two rollers. In case of mechanical impact, the feedstock is compressed between the units of grinding media, or grinding media and the mill lining (though the latter is not desired), and experiences a shear force. In particle attrition, wear from the surface of the feedstock particles leads to the reduction in size of the primary particles and generation of secondary, very fine particles. It will be, however, seen below that the experimental set-ups in the two cases can have certain similarities in the basic arrangements, e.g. in ordinary ball mill and attrition mill. For natural and synthetic ceramic powders three types of grinders or mills are generally used : simple ball mills, vibratory mills and attrition mills or attritors. In all cases, the milling process needs grinding media, and may take place under dry or wet conditions. They thus distinguish themselves from other kinds of mill, e.g. roller mills and jet mills which do not require any grinding media. The approximate relative performances of some of the mills are shown in Fig. 3.1.

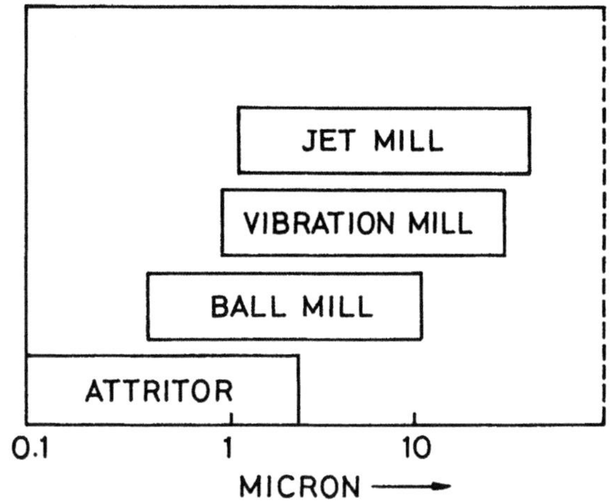

Figure 3.1 Approximate relative performance ranges of some mills.

The oldest and most popular industrial means of comminution of relatively small grains into generally micron-sized particles is the ordinary (or tumbling) ball mill which functions mainly by compression and shear action (friction) on the charge, as also by mechanical impact. A common laboratory set-up (Fig. 3. 2) consists of

(i) a cylindrical chamber (also called a "pot" or a "mill") made of or lined with porcelain, alumina, zirconia, tungsten carbide etc. rotating horizontally on its own

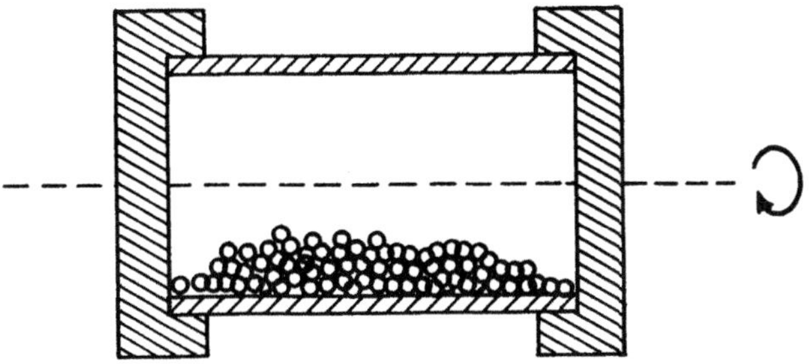

Figure 3.2 Schematic representation of a common laboratory ball mill.

axis, and (ii) hard ceramic or metal grinding media (often of the same material as that of the chamber lining) of spherical, cylindrical, i.e. rod-like or irregular but rounded shape. The spherical (ball-like) media give the equipment its name. Due to its long-standing acceptance in the laboratory and the industry, an ordinary ball mill comes in all sizes, with a capacity of several grams to several tons.

The feedstock batch is charged into the chamber (occupying about 50% of the space), maintaining fixed weight ratios of charge : grinding media : water (the latter when wet milling is necessary and permitted). Slow horizontal rotation (commonly below 100 rpm) of the chamber causes progressive reduction in size of the feedstock material via interaction with the media. Simple calculations [126] show that the critical rotational speed v required to take the media to the top of the horizontally rotating chamber would depend only on the size of the chamber (R) :

$$v = (g/R)^{1/2} \qquad \text{... (3.1)}$$

where g = acceleration due to gravity.

Thus, a chamber with 150 mm diameter can be run with a speed of nearly 700 rpm. The nature of movement of the media as a function of the speed of rotation has been schematically depicted in Fig 3.3 (a) and 3.3 (b). Depending on the capacity of the mill, size of the charge etc., the grinding media can have different sizes [125].

Impurity pick-up is always a big concern in the milling of relatively pure powders; to avoid or minimize this during the milling of high-purity powders, the chamber and the media are generally made of the same composition as that of the feedstock material, or harder substances. The grinding media are expected to be made of not only theoretically hard substances, but also high-density, sintered bodies, so that the impurity pick-up can be minimized. However, the high processing cost of

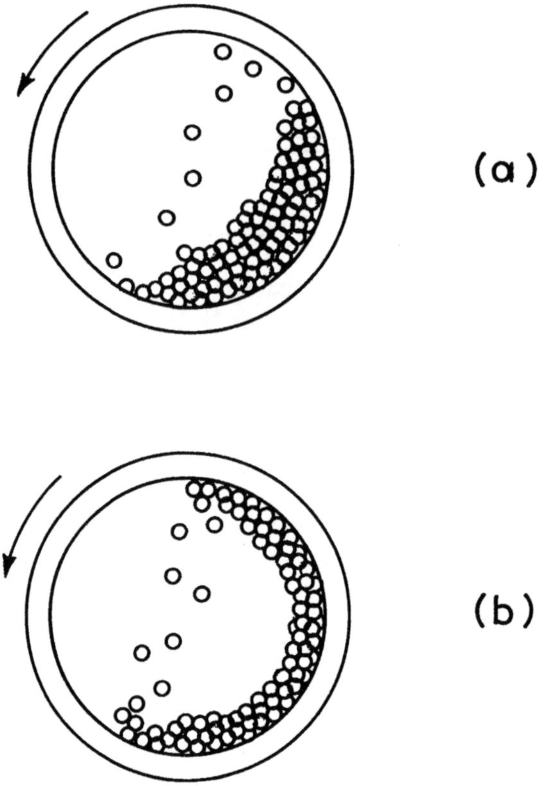

Figure 3.3 Schematic representation of the movement of the grinding media in a ball mill : (a) low speed; (b) high speed.

obtaining sintered media or chamber lining of close to theoretical density often precludes their use in industrial milling operations (leading to the generation of impure milled powders). This is an important reason why hard agglomerates should be avoided in powder production, so that chemical contamination through milling can also be avoided. So far as the chamber lining is concerned, an alternative is the use of materials like soft rubber or polyurethane, shreds of which get mixed with the powder under grinding, but can be burnt out later. Again, residual carbonaceous materials can cause microstructural defects in the sintered products fabricated out of such powders.

Wet milling has the advantage of producing finer particles than dry milling, and is also known for greater utilization of the energy which is linked with the economics of the operation. However, there are also several limitations of wet milling :

(i) Wear of the grinding media can be higher, hence the powder can become more contaminated.

(ii) The milled particles may require additional steps like washing, sedimentation etc. for removing the contaminants.

(iii) The milled particles require drying, which involves additional expenditure of energy.

(iv) Some particles may release surface ions (e.g. nitrogen in AlN) by reaction with water, and hence, cannot be used as such.

An attrition mill, also called an "attritor", a "stirred ball mill" or an "agitated ball mill", is supposed to be the most efficient among the grinding media-based compression-shear type mills. It usually consists of a *stationary* chamber housing a centrally positioned rotating stirrer system, with several arms attached to a central shaft at right angles (agitator system) and the necessary grinding media. The chamber can be vertical or horizontal. The feedstock (mostly slurries, but also dry powder), transferred into the chamber by simple charging or pumping, is agitated along with the media by the agitator system and is reduced in size through wear from the surface, i.e. very fine particles are continuously generated. Figure 3.4 shows the basics of a laboratory attrition mill.

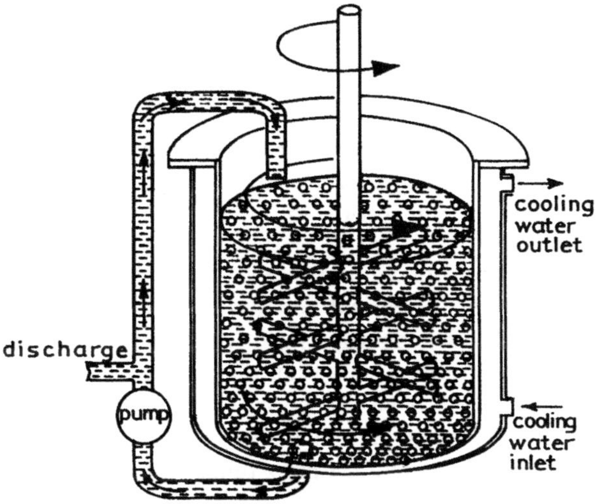

Figure 3.4 Inside view of a laboratory attrition mill
(kind courtesy : Union Process, Akron, Ohio, U.S.A.).

The importance of attrition milling lies in the fact that the power consumed by the equipment is spent directly in agitating the media, and not in causing movements of the milling chamber, as in tumbling ball mills. The energy utilization of the equipment is thus relatively high. The stirring system rotates at a frequency of 1–10 Hz, and the tips of the arms can achieve a speed of $18 \times 10^3 - 30 \times 10^3$ cm/min or higher. Due to the high speed, the grinding media can be small, i.e. < 2 – 10 mm beads and balls; they are commonly made of alumina, zirconia, silicon carbide and nitride etc. [127].

An attrition mill, first conceptualized by Andrew Szegvari in the 1920's [128], can be of three types : batch-, continuous- and circulation-type. Apart from the energy economy, attrition milling offers the following (some of them related with the energy utilization) advantages [127] :

(i) Fast and efficient fine grinding (down to ~ 0.1 μm) can be achieved.

(ii) Less contamination from grinding media and chamber wall is expected.

(iii) Simple operation.

(iv) Efficient temperature control (all grinding tanks have cooling or heating jackets).

(v) Small area requirements.

The limitations are :

(i) The usefulness is almost restricted to fine feedstock and grinding range.

(ii) Wet milling is almost a routine requirement (as it produces finer particles than dry milling), the disadvantages of which have been described earlier in this Section.

(iii) Continuous operation generates considerable heat, requiring a cooling system; this is of course generally provided in all systems (see above). Figure 3.5 shows a large attritor for industrial operations.

A vibratory (attrition) mill or vibro-mill as it is commonly known consists of a rapidly vibrated chamber in vertical position which is nearly completely filled (~ 90% of the volume) with generally cylindrical media and the slurry of the feedstock (mills with low-amplitude vibration are generally used for wet grinding). The frequency of vibration is so chosen as to be in harmonic motion with the natural frequency of the equipment, commonly 10–30 Hz [129]. The media are made of alumina, zirconia, silicon carbide, tungsten carbide and steel.

The three-dimensional vibration pattern is controlled so as not to disturb the packing arrangement of the grinding media and to control the energy requirement. The impact energy generated in the process is much greater than that in an ordinary ball mill, and so, the process is more energy efficient than ordinary ball milling.

Of the two major milling processes that do not require grinding media, high-compression roller milling provides one example. It is usually meant for milling of relatively large chunks down to ~ 10 μm only. In essence, the system (Fig. 3.6)

Figure 3.5 Attritor for industrial operations
(kind courtesy : Union Process, Akron, Ohio, U.S.A.).

consists of two oppositely rotating rollers between which there is a controlled space for the charge to pass through (under gravity) at a high contact pressure. The process is energy efficient, but for obtaining fine ($\leq 1\ \mu m$) particles, this mill must be used in conjunction with, say, an ordinary ball mill.

Jet milling or fluid energy milling, the other method that works without grinding media, offers a very interesting way of comminution; developed in the 1880's, such mills are still an important means of size reduction, and are available in different models. Generally, jet milling is done by the interaction of high speed

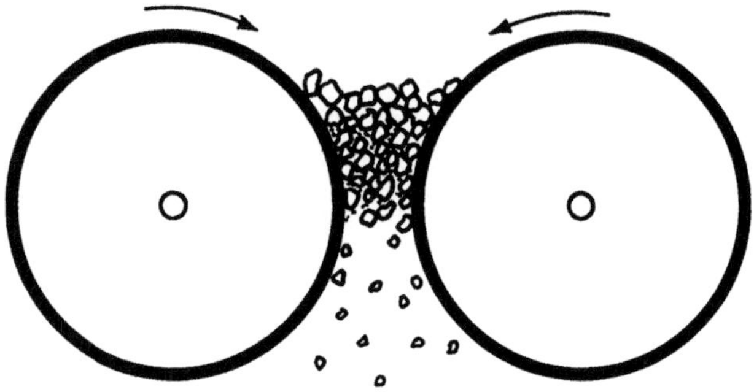

Figure 3.6 Schematic representation of a roller mill.

gas streams entering the milling chamber through nozzles placed at suitable angles, at least one of them carrying the feedstock particles. The particles collide with each other at a high speed in a turbulent zone inside the chamber, and in the process, undergo fracture and comminution into smaller size (~ 1 μm). The gases commonly used are steam or air under a pressure of ~ 20×10^5 Pa and ~ 7×10^5 Pa respectively [129]. Inert gases like N_2 or Ar can also be used according to necessity.

Gas streams coming out of the chamber carry the fine-milled powder, which is then collected in a cyclone chamber. Jet milling usually operates by the collision (and re-collision) of feed particles and size classification, but in some models the particles also hit the chamber lining. Obviously, little contamination of the milled powder is usually expected from this operation. However, when a particle-lining collision is envisaged, the lining is made of relatively soft substances.

3.2 SOLID–SOLID REACTIONS

3.2.1 Mixing and Calcination

Solid-solid reactions by conventional mixing and calcination pertain to the synthesis of a compound through the reaction between/among two or more starting materials commonly at atmospheric pressure and elevated temperatures. Two major processes are involved in solid-solid reactions :

(i) Breaking and reconstruction of bonds at the contact region, leading to the nucleation of a product phase.

(ii) Transport of matter to the contact region.

In practical situations the process of transport is slower, and hence, the rate-controlling step. The transport of a reacting species (e.g. the cations A^{2+} or B^{4+} in Fig 3.7) is described by its rate of diffusion, i.e. the diffusion coefficient. Depending upon the stage and mode of reaction, diffusion may take place at the surface or in the bulk.

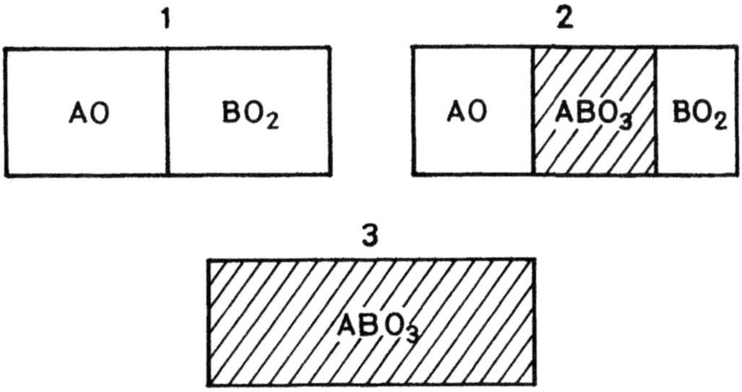

Figure 3.7 Schematic of solid-solid reactions : (1) reactants (AO and BO_2) in contact at zero time; (2) intermediate stage of reaction indicating partial formation of the product (ABO_3); (3) product after completion of reaction.

Considering steady-state diffusion, one can use the well-known expression

$$D \text{ (div grad) } C = 0 \qquad \text{... (3.2)}$$

where D = diffusion coefficient and C is the concentration of the diffused species in the matrix in which diffusion has taken place.

When the diffusion is unidirectional, the growth rate of the product layer is indicated by the relationship

$$dx/dt = Dk/x = k_1/x \qquad \text{... (3.3)}$$

where x = thickness of the product layer, t = reaction time, D = diffusion coefficient and k is a proportionality constant. When the diffusion is independent of time and the contact area between the two reacting particles remains constant, we can derive from the above

$$x^2 = 2kDT + C \qquad \text{... (3.4)}$$

At the initial stage, i.e. at $t = 0, x = 0$,

$$x^2 = 2kDt = k_p t \qquad \text{... (3.5)}$$

where k_p is the parabolic rate constant. The relationship is known as the parabolic rate law which indicates, at least theoretically, how a diffusion-controlled solid-solid reaction is likely to proceed.

Ceramists preparing powders via solid-solid reactions agree that except in model experiments with powder compacts, it is impossible to obtain a realistic estimate of x, the thickness of the product layer. Instead, it would obviously be much simpler to determine the residual volume of the reactants remaining unconsumed during the formation of the product after time t. W. Jander [130] considered a situation where particles of one reactant were completely and constantly covered with those of another reactant (i.e. stoichiometric excess of the latter over the former); making a large number of assumptions, e.g. the reactant particles are spherical, the reaction is controlled by (unidirectional) bulk diffusion, the parabolic rate law is followed, the temperature is fixed etc., he proposed a kinetic model [131]. A simple and usable relationship that can be derived from the model is :

$$k_p t = r^2 \, [1 - (1 - y)^{1/3}]^2 \qquad \text{... (3.6)}$$

$$\text{or,} \quad k_j t = [1 - (1 - y)^{1/3}]^2 \qquad \text{... (3.7)}$$

where k_j is the rate constant and y is the fraction of reaction completed in time t.

The above relationship derives from the following. When the volume of unreacted material at time t is V, then

$$V = \frac{4\pi}{3}(r - x)^3 \qquad \text{... (3.8)}$$

where r is the initial radius of the reacting particles (uniform spheres) and x is the thickness of the product layer. When y is the fraction of the reaction completed at time t,

$$V = \frac{4\pi r^3}{3}(1 - y) \qquad \text{... (3.9)}$$

Eqns. 3.8 and 3.9 yield

$$x = r_0 \, [1 - (1 - y)^{1/3}] \qquad \text{... (3.10)}$$

assuming that the reaction would follow the parabolic rate law (Eqn. 3.5).

In spite of the fact that the above retationships are based on a large number of assumptions (e.g. very few varieties of commercial powders are uniformly spherical), the parabolic reaction rate has been found to hold good for various reactions involving solids, especially at the early stages. An improved relationship was proposed by Carter [132, 133] :

$$[1 + (z - 1)x]^{2/3} + (z - 1)(1 - x)^{2/3} = z + 2(1 - z)\,\frac{kt}{r^2} \qquad \text{... (3.11)}$$

where z = volume of the product obtained from a unit volume of reactant composed of particles of radius r and x = relative amount of the reactant converted to the product.

In practical ceramic powder preparation, the reactants are generally crystalline metal oxides, hydrated oxides, hydroxides, carbonates etc. When one of the reactants is a hydroxide or a salt, the following are to be considered :

(i) At elevated temperatures of calcination, the hydroxide or the salt will decompose, yielding a gas phase. This may react with the solid reactant phases, and the system may not represent an all-solid reaction any more. Further, the partial pressure of the evolved gas in the reaction atmosphere (generally air) may influence the reaction kinetics.

(ii) At the decomposition temperature, there is a drastic structural change involving bond breaking; the particulate product is also usually finely divided. The decomposition temperature is therefore one at which a high reactivity is expected. This is known as the "Hedvall effect", named after the Swedish chemist, and is taken advantage of in solid-solid reactions. The effect is also valid at the temperatures of polymorphic phase transformations of crystalline reactants.

3.2.2 Modern Techniques

The basic principle of comminution has been utilized in recent times for causing solid-solid reactions within shorter time periods than those in the conventional methods of mixing and calcination. Two such methods will be described below.

High energy ball milling has been demonstrated to be a technique for the synthesis of known or new compounds by solid state (including solid-solid) reactions. Often called "mechanochemical synthesis", it is carried out by feeding the charge into a cylindrical vial containing the milling media in required proportions. Vibratory, planetary or ordinary ball mills are used with tungsten carbide, alumina (sapphire), hardened steel or other kinds of hard vials and matching grinding media [134–136]. The charge (e.g. graphite and silicon for SiC), during milling, undergoes three-dimensional, large-amplitude vibrations commonly at a frequency of ~ 20 Hz [137]. The powder charge : grinding media weight ratio varies in the range 1 : 10 to 1 : 2 according to necessity; the time of milling is generally 24-30 h. Note that all reactions take place at nominally room temperature (i.e. not considering the frictional heat). High energy ball milling is an interesting technique for the synthesis of carbides and silicides [134–136]. In addition, recent work [138] reports borothermic reduction of TiO_2 and ZrO_2 to the respective borides.

Microwaves, which are coherent and polarized electromagnetic waves with the frequency range 0.3–300 GHz (~ 1 m–1 mm wave length) are now being used for not only drying or sintering but also for the synthesis of compounds of ceramic interest. Briefly, the syntheses are based on the fact that above a critical temperature, or even at room temperature, some substances can absorb microwaves and couple with them, so as to obtain the energy required for reacting among themselves [139, 140].

One of the most attractive features is the relatively very short time required to complete the reaction and obtain a nearly phase-pure compound. However, to achieve this, the container housing the reactant should preferably not absorb microwaves. Preparation of ultrafine mullite and superconductor powders by microwave-assisted reaction was reported during the early experimentations [139]. In a later review of the activities on microwave synthesis, Mingos [141] reported the preparation of aluminosilicate zeolites in 2–3 mins by using a microwave power of 300 W. Similarly, yttrium-barium-copper oxides have been prepared within 1.1 h as compared to the conventional requirement of 24 h.

Among the non-oxides, β-SiC powder has been synthesized in a commercial microwave oven operating at 2.45 GHz and a power of up to 980W [142]. Aluminium nitride powder has also been synthesized recently by the reaction between Al metal and carbon in a nitrogen atmosphere; the recent advances in microwave synthesis have been reviewed by Rao and Ramesh [143].

On the face of it, the carbothermal reduction processes for the synthesis of carbides can be considered to be solid-solid reactions, e.g. $SiO_2 + 3C \rightarrow SiC + 2CO$. However, as the reduction does not take place without the presence of N_2, Ar etc., they have been considered in this text to be vapour-solid reactions.

3.3 SOLUTION TECHNIQUES

3.3.1 Precipitation and Co-precipitation

Precipitation of a solute (commonly an inorganic or organic salt) from a homogeneous solution as a function of its concentration in the solvent and factors like pH, temperature etc. is a well-known process in industrial chemistry for obtaining crystalline particles or grains of various inorganic and organic compounds [144]. The solubility of the salt in the solvent under a given set of conditions is the maximum quantity that can remain dissolved in a unit volume of the solvent, i.e. the saturation limit up to which the solvent holds the solute in a dissolved state. The equilibrium saturation concentration, c^* is given by the empirical relations of the type

$$c^* = s_0 + s_1\theta + s_2\theta^2 \qquad \text{... (3.12)}$$

where θ is the solution temperature and s_0, s_1.... are the coefficients of the components [145]. A supersaturated solution (where the concentration exceeds the saturation limit) tries to obey the equilibrium conditions by precipitating out the dissolved solute in the form of a solid of the same composition. Crystalline salts obtained in this way are often thermally decomposed to convert them to oxides of ceramic interest.

The precipitation technique as a whole is very well-known as a commercial process of manufacturing ceramic powders, though the involved procedure and chemistry may be different from that discussed above. The starting materials, as above, are in general salts of strong acids which can be represented by, say, $A^{x+}(NO_3)_x$, i.e. a nitrate salt, containing the cation A^{x+} which is required to be precipitated as a part of a solid, e.g. the corresponding hydroxide. The salt in the dissolved state in water dissociates into A^{x+} and $(NO_3)^-$ ions, both surrounded by the water molecules, i.e. in a hydrated state if permitted by the basicity of the system. This is so, because a cation dissolved in water can also progressively hydrolyze [146] in a pH dependent way by reacting with it.

$$A^{x+} + yH_2O \leftrightharpoons A(OH)_y^{(x-y)+} + yH^+ \qquad \text{... (3.13)}$$

A general expression can be written as

$$zM^{x+} + yH_2O = M_z(OH)_y^{(xz-y)} + yH^+ \qquad \text{... (3.14)}$$

The hydrolysis can occur at even very low pH if the charge/diameter ratio is high (as in the case of Si^{4+}). The hydrolyzed product, as shown above, can grow in size with the degree of hydrolysis, as also go through a process of condensation :

$$M\text{–}OH + M\text{–}OH \rightarrow M\text{–}\overset{\overset{\displaystyle H\ \ OH}{|\quad |}}{O\text{–}M} \rightarrow M\text{–}O\text{–}M + H_2O \qquad \text{... (3.15)}$$

to form polynuclear species; polymerization of the species helps further growth. For cations of large size and small charge (e.g. Pb^{2+}), hydrolysis (followed by condensation) takes place at a relatively high pH. For causing rapid hydrolysis and facilitating a quick growth of hydrolyzed-condensed species of most metal ions, a rise in the basicity of the system is necessary. This is provided by the contact of a metal salt solution with solutions like NH_4OH at, say, pH > 7. The derived metal species then precipitate in the form of (commonly amorphous) hydroxides or hydrated oxides. In the precipitation reaction, aqueous salt solutions of metal nitrates, chlorides or sulphates are used. For a multicationic solution, intermediate compound formation and precipitation must not occur by the exchange of the ions, i.e. the starting salts must be system-compatible to obtain a homogeneous solution.

The precipitants commonly selected are NH_4OH, NaOH, Na_2CO_3 or a suitable mixture of NaOH–Na_2CO_3 or NH_4OH–$(NH_4)_2CO_3$. The addition of either the precipitating agent to the salt solution or vice versa, i.e. the sequence of addition, is an important point to be considered. When a single cation is present in the starting solution, the precipitant, usually NH_4OH, is poured in excess into it to raise the solution pH to more than the desired value and to obtain the precipitate. This is called the "direct strike" method. When two or more cations are present in the solution, concurrent and homogeneous precipitation of the corresponding two or more hydrated species becomes difficult. The main reason is that, as discussed above, the hydrolysis of different cations requires different basicities, i.e. pH values of the solution [146]. To avoid stepwise and local precipitation (leading to a gross inhomogeneity in the final product), the starting mixed solution is added in a stream into an excess of NH_4OH solution. This is called the "reverse strike" method. Such a process is termed, though not in a strict sense, "co-precipitation".

The particle size of a precipitate decreases with increasing concentration of the metal ions in the solution. This is caused by the fact that an increase in the concentration causes a high degree of supersaturation with the formation of a large number of nuclei. On the other hand, a dilute solution favours the formation of comparatively large particles due to a low degree of supersaturation. The precipitation reaction is also affected by the temperature of the salt solution. The solution is generally heated at a temperature below its boiling point, [e.g. 147]. At an elevated temperature, the solubility increases, and consequently, the supersaturation decreases. Further, with the increase in temperature, both the hydrolysis and polymerization reactions are facilitated.

The final pH of the solution is also important in the precipitation reaction. For compounds which exist in several polymorphic forms, e.g. ZrO_2, the final pH of the system will determine the type of the polymorph crystallizing (on calcination) out of the precipitate obtained at that pH [148]. At a low pH, incomplete precipitation can occur due to an insufficient supply of OH^- ions. On the other hand, excess OH^- ions can form a water-soluble complex in some cases with the metal ion, making the precipitation unfeasible [149].

As mentioned earlier, in addition to the precipitation of hydroxides or hydrated oxides, that of salts like oxalates is also related to the pH, with oxalic acid or ammonium oxalate as precipitants; the relative amounts of $H_2C_2O_4$, $HC_2O_4^-$ and $C_2O_4^{2-}$ formed depend mainly on the pH of the solution. Further, the pH value also controls the oxidation state of the metal ion, e.g. in presence of atmospheric oxygen and at high pH, Fe^{2+} transforms to Fe^{3+} [149].

The avoidance of agglomeration is a very intricate and important problem for powders precipitating and precipitated from solutions. Various means have been adopted and recommended by workers in this area [e.g. 4], but unfortunately very few commercial or even research powders are reported to be absolutely agglomerate-free. Some of the procedures (see also Table 1.1.) are listed below.

(i) Vigorous stirring of the solution during and after precipitation.

(ii) Spraying or atomization of (mixed) solutions on to the surface of a pool of ammonia solution under stirring, keeping the pH of the system at a constant value.

(iii) Hydrolysis of the metal in solution with ammonia gas introduced to the liquid surface, using an inert gas, e.g. argon as the carrier. The liquid-gas interface reaction develops weakly agglomerated precipitates.

(iv) Addition of H_2O_2 to the salt solution. Hydrogen peroxide forms peroxo complexes with the metal ions, e.g. a titanyl complex, stable in an aqueous solution, in case of Ti^{4+} [150]; this suppresses the hydrolysis of the metal ions. Precipitation in such cases is carried out at a temperature somewhat higher than the ambient, say, 40°C.

(v) Removal of the anions of the salt solutions used, as they may help the formation of salt crystals joining individual particles. Repeated washing ensures the removal to a great extent, but in the process causes partial peptization (breaking down of the particles in the precipitate). This can cause the formation of hard agglomerates during calcination.

(vi) Washing of the precipitate with organic solvents of lower surface tension than that of water. Methanol, ethanol, acetone, isopropanol, acetone-toluene-acetone (in a sequential mode) etc. are often used for washing. Low surface tension of the pore liquid in the precipitate reduces the capillary force, producing soft agglomerates after calcination.

(vii) Drying of the precipitate at high humidity (~ 95%) and temperature (~ 90°C) can produce agglomerates which are easily crushable.

(viii) Fine crystallite growth by hydrothermal treatment of the amorphous precipitate (Section 3.3.3).

(ix) Freeze-drying of the precipitate (Section 3.4.4).

3.3.2 Forced Hydrolysis

As discussed in Section 3.3.1, metal cations in an aqueous solution are known to hydrolyze under specific chemical conditions, especially the pH of the solution. The hydrolyzed cations form hydroxides or hydrated oxide species and precipitate as uncontrolled agglomerates by supersaturation unless they give rise to only one burst of nuclei which grow by a diffusion process. To obtain monodispersed, precipitated particles of a narrow size distribution, the hydrolysis reaction is carried out in a controlled way, so that only a small degree of supersaturation is achieved in the solution. Under such conditions, the mechanism of precipitation is best represented as [79]

$$n[M(H_2O)_p]^{z+} \xrightarrow{OH^-} [M_n(H_2O)_{np-m}(OH)_m]^{(nz-m)+} \xrightarrow{OH^-} nM(OH)_z \qquad \text{... (3.16)}$$

The intermediate complexes thus formed act as the precursors of the nuclei which subsequently grow in size during the course of the reaction. The physical and chemical characteristics of the ultimate precipitate are governed by the composition as well as the rate of formation of intermediate precursor complexes. Such controlled hydroxylation of the metal ions cannot be effected reproducibility by the direct addition of a base to the salt solution, as the precipitation reaction depends on a number of parameters discussed in Section 3.1.1.

A slow addition under stirring of a precipitant of low concentration to a hot salt solution favours a low degree of supersaturation. Further, hydrolysis of a salt solution is known to be facilitated at high temperature. Such *forced* hydrolysis of the metal ion liberates proton-producing complex species as precursors. The liberation of proton is controlled by controlling the ageing time at a fixed temperature and metal ion concentration. Under such conditions, controlled growth of the nuclei occurs by diffusion, forming uniformly sized particles (one burst of nuclei [79]).

In actual practice, this controlled hydrolysis is conducted by ageing an acidified salt solution at an elevated temperature, thus forcing the hydrolysis reaction to take place within definite times. The OH^- ion necessary for metal hydroxylation is produced in situ by the deprotonation of the water molecules attached to the metal ion. Homogeneous precipitation is also caused by the slow release of ammonia through the hydrolysis of urea, $CO(NH_2)_2$ present in the precursor solution at ~ 70°C [151].

The reactions involved are :

$$CO(NH_2)_2 \leftrightharpoons NH_3 + HNCO \leftrightharpoons NH_4^+ + NCO^- \qquad \ldots (3.17)$$

In an acidified solution (as mentioned above)

$$NCO^- + 2H^+ + H_2O \rightarrow CO_2 + NH_4^+ \qquad \ldots (3.18)$$

The dissociation constant of urea in water being very low ($K_b = 1.5 \times 10^{-14}$), the rate of hydrolysis can be easily controlled. Hydrolysis can be stopped at a required pH by cooling the system to the ambient temperature.

The precipitate obtained by this method is generally dense and can be easily filtered. In general, a slow reaction favours the formation of particles of large size, which assists easy filtration. The morphology of the particles obtained by the homogeneous precipitation reaction can be varied dramatically (spherical, cubic, acicular etc.) by changing the rate of the reaction, type of the anion in the starting salt (sulphate, chloride etc.), temperature of ageing and also the final pH of the solution. Matijevic and co-workers have pioneering and extensive contributions [e.g. 79, 151] in this area which show the versatility of the technique and the contribution of the different parameters to the properties of the final product, especially the particle shape. The morphology as a function of the anion has also been studied by various other workers [e.g. 152].

The powders obtained by this process can be (i) collected with relative ease and washed in non-aqueous media (e.g. ethanol) to avoid agglomeration or (ii) obtained as colloids in a homogeneous sol.

3.3.3 Hydrothermal Synthesis

The hydrothermal technique of powder preparation involves reactions taking place in an aqueous or aquo-organic environment, but at relatively high temperatures and pressures with respect to the ambient conditions. The hydrothermal fluid environment, thus, is created in a pressurized and heated sealed system (e.g. in a conventional pressure vessel made of hardened steel or other alloys), where the reactants are housed. They may be particles of metal, oxide, hydroxide etc. or amorphous precipitates and gel fragments, as also different salt solutions, some of which play the role of catalysts. The temperature and pressure can vary in the range 80° – 400°C (in some cases even higher) and small autogenous pressures (that obtained only by the volume expansion of the fluids at the above temperatures) to 100 MPa (or more) respectively. Noble metal liners or "capsules" (open or closed) have often been used inside the pressure vessels, e.g. to keep the reactants insulated from impurities; polymer linings are used for resisting the corrosion of the vessel.

Somiya et al. [153] have classified hydrothermal reactions yielding fine powders into five groups :

(i) Oxidation, e.g. $Zr + H_2O \rightarrow ZrO_2 + H_2$.

(ii) Precipitation, e.g. $KF + MnCl_2 \rightarrow KMnF_3$.

(iii) Synthesis, e.g. $La_2O_3 + Fe_2O_3 + SrCl_2 \rightarrow$ (La, Sr) FeO_3.

(iv) Decomposition, e.g. $FeTiO_3 \rightarrow$ Fe-oxide + TiO_2.

(v) Crystallization, e.g. hydrous Zr-oxide $\rightarrow ZrO_2$ (t- or m-phase).

It is apparent that practically all of the above variants of the hydrothermal technique are important for the preparation of ceramic powders, but a large majority of the work has been done by using the oxidation, crystallization and synthesis modes. Both oxidation and crystallization modes have been found to be suitable for the generation of nanoparticles of high temperature oxides like ZrO_2, while the decomposition mode relates to, e.g. the conversion of $ZrSiO_4$ to ZrO_2. The synthesis mode was also found suitable for titanates and zirconates, e.g. $BaTiO_3$ and lead zirconate-titanate (Chapter 4).

Powders obtained by the hydrothermal technique exhibit characteristics which make them ideal as precursors for sintered ceramic products [154] : (i) fine particle size, i.e. 10–100 nm, but often 10–30 nm; (ii) narrow particle size distribution; (iii) soft agglomerates, or agglomerate-free; (iv) chemically pure and homogeneous; (v) well-crystallized (often single crystal) particles of characteristic morphology; (vi) low temperature product; (vii) relatively strain-free.

Zirconia is one of the compounds which have been considered for commercial powder production by the hydrothermal technique. The suitable methods have been discussed below in brief.

The hydrothermal oxidation process has been referred to as a breakdown method [155] because the reactions involved in the process divide the grains into impurity-free particles of nanometer size, e.g. 20–30 nm. Reaction between high purity Zr metal and pure water in a closed vessel at up to 700°C and under a pressure of 100 MPa may be considered as an example. For the preparation of stabilized ZrO_2, salt solutions of Ca or Y (as desired) can also be used in place of pure water.

$$Zr + 2H_2O \xrightarrow{300^\circ C} m\text{-}ZrO_2 + ZrH_2 + Zr \text{ (unreacted)};$$

$$Zr + H_2 \rightarrow ZrH_2 \qquad \text{... (3.19)}$$

$$ZrH_2 + 2H_2O \xrightarrow{>400^\circ C} ZrO_2 + 3H_2 \qquad \text{... (3.20)}$$

As against the above, the hydrothermal crystallization technique is called the build-up method. Again, the hydrothermal crystallization of zirconia can be taken as an example. Amorphous hydrated zirconia is obtained by precipitating $ZrCl_4$ solution with NH_4OH followed by washing with distilled water and drying at 120°C for 48 h. This starting material in presence of a mineralizer, e.g. KF solution, is known to crystallize at 200° – 600°C for 24 h at a pressure of 100 MPa. The product is monoclinic zirconia nanoparticles, e.g. 16 nm at 200°C and 22 nm at 500°C.

Hydrothermal decomposition of zircon is known to occur in presence of NaOH and $Ca(OH)_2$. Typical experimental conditions are [156] :

NaOH concentration in the solution	: 7 wt%
Molar ratio of $Ca(OH)_2$ / $ZrSiO_4$	: 2
Temperature	: 350°C
Pressure	: 17 MPa
Time of reaction	: 8 h

Zircon reacts with $Ca(OH)_2$, with NaOH acting as a catalyst :

$$ZrSiO_4 + xCa(OH)_2 \rightarrow ZrO_2 + xCaO.SiO_2.H_2O + (x-1)\, H_2O \qquad \text{... (3.21)}$$

On the other hand, the hydrothermal reaction between zircon and NaOH can be depicted as

$$ZrSiO_4 + 2NaOH \rightarrow Na_2ZrSiO_5 + H_2O \qquad \text{... (3.22)}$$

When both NaOH and $Ca(OH)_2$ are present, Na_2ZrSiO_5 reacts with the reaction products of Eqn 3.21.

The solid products are removed by filtration and washing, and acid-leached to obtain a powder batch of the hydrated oxide of zirconium, which is washed and calcined.

3.3.4 The Sol-Gel Process

Various processes of synthesis of single or multiple oxide powders are recorded in the literature which start with the preparation of an aqueous or organo-sol. The subsequent processing steps vary quite significantly, but the basic chemical principles may remain the same. As a result, it is difficult to describe a single and unique powder preparation technique as the "sol-gel process". As will be apparent to the reader, some of the techniques described elsewhere in this Chapter have been utilized as successfully with a sol as a solution or slurry as the starting material. In this discussion, we consider those "sol-gel" processes which use bulk gels as solid precursors for powder preparation.

Before we proceed further, it seems in order to discuss briefly the basic traits of a sol, some techniques of preparation of a sol, and some aspects of sol-gel conversion. In very broad terms, a sol is defined as a stable dispersion of colloidal particles (up to about 1 μm) in an aqueous, aquo-organic or organic liquid medium. A sol is generally prepared by one of the three main ways :

(i) by dispersing a batch of particles of the desired size range in a liquid medium;
(ii) by causing nucleation and growth of particles of the desired size range within the solution; and
(iii) by precipitating large particles and agglomerates in the liquid by, e.g. hydrolysis reaction of a premixed precursor solution, and peptizing [157] them into much smaller particles of the desired size range, so that they form a stable dispersion (see below).

In the method (i) above, two points are to be taken into consideration : the particles must be wetted well by the liquid and must possess the same surface charge so that a repulsive force acts among them. These concepts have been introduced briefly in the earlier Chapters; further reference may be made to Chapter 1.

The method (ii) of generating particles in a liquid medium generally involves the control of the hydrolysis-polycondensation reaction of metal-organics (e.g. alkoxides, modified alkoxides) or metal salts. Reactions involving the hydrolysis of alkoxides, more than any other reaction, have been the hallmark of the modern sol-gel chemistry [158]. The hydrolysis reaction generally involves controlled addition of water in an alcoholic solution of a metal alkoxide :

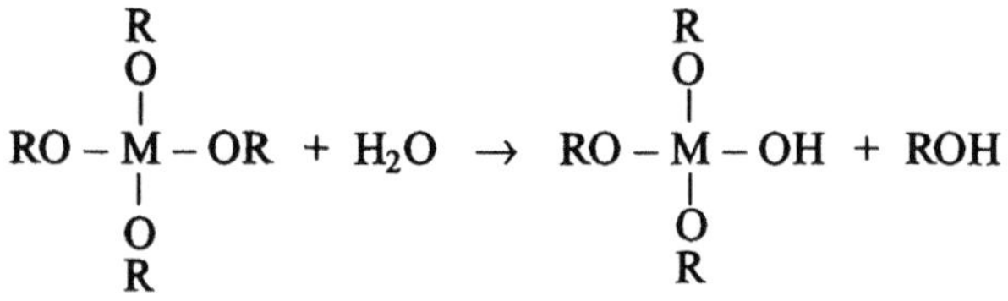

... (3.23)

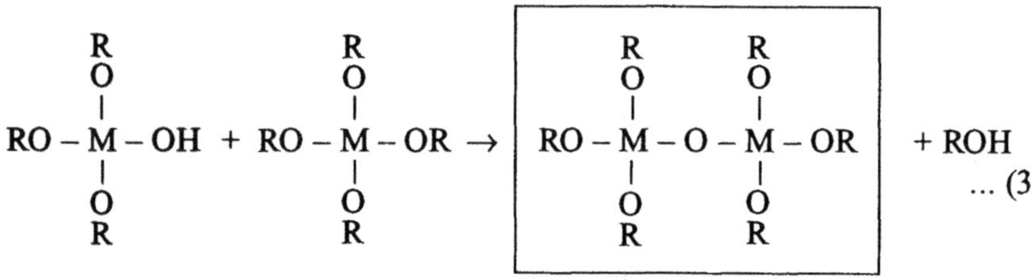

... (3.24)

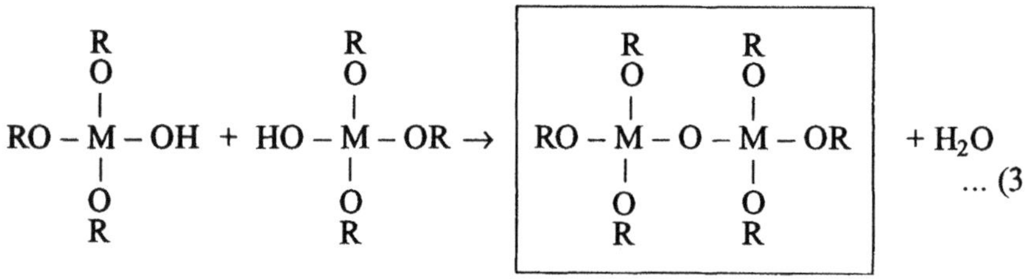

... (3.25)

where R = CH_3, C_2H_5, C_3H_7 etc.
M = Metal cation (e.g. Ti^{4+}, Si^{4+}), in this case in four-coordination with the OR groups.

Reaction 3.23 indicates the initiation of hydrolysis. The primary reaction products within the cages in Eqns. 3.24 and 3.25, exhibiting M – O – M bonds, indicate the initiation of condensation polymerization [158].

A progress of these reactions causes further polymerization and (usually) three-dimensional growth of these branched polymers into "particles". The product thus obtained matches the definition of a sol, and is often called a "polymeric" sol. Metal cations, dissociated from inorganic or organic salts dissolved in water, become hydrated at low pH, but are hydrolyzed with increase in the pH, the critical value of which depends on factors like the charge and size of the cation (Section 3.3.1) [146].

The hydrolyzed products polymerize and condense to form relatively large polynuclear species which can be termed "particles" as in the case of alkoxide-derived polymers at an early stage of linking.

The method (iii) takes the product of uncontrolled precipitation of the hydrated oxide or hydroxide of a metal ion in an aqueous medium as the starting material for obtaining a sol. Such flocculated precipitates are washed free from the electrolytes and the base used for precipitation. The next step is the breaking of the precipitate into smaller units by the action of acid, i.e. by the adsorption of H^+ on the particle agglomerate surfaces, setting up repulsive forces and causing their breakdown into smaller particles. The other phenomenon that takes place (depending on the pH conditions) is the surface dissolution of the particles, making them smaller in dimensions. The overall process is generally described as *peptization*. The small particles thus produced become dispersed in the liquid, and form a sol.

For obtaining solid matter from a stable sol, it must first be destabilized. There are several methods of achieving this. In essence, all these methods involve a progressive reduction in the distance between the suspended particles and finally, the formation of a network structure through contact among them.

The available liquid phase is located within the pores of this network which is called a gel. A gel is thus a rigid to semi-rigid porous solid, and its generation from a sol is called the sol-gel transition. Under controlled conditions of transition, a gel is obtained as a monolithic body which takes after the shape of the sol, i.e. that of the container housing the sol.

Monolithic gels (aquo- or alcogels depending upon the preponderant liquid, i.e. water or alcohol in the pores) are difficult to dry or heat without sacrificing the monolithicity due to the capillary pressure [158]. Uncontrolled drying or heating shatters them to small pieces.

Bulk gels can be used for powder preparation in different ways, e.g. :

(i) The gel is dried under high vacuum, followed by grinding and sieving to obtain the precursor gel powder. Calcination of the gel powder at the desired temperature produces a crystalline oxide powder.

(ii) The gelled mass is dried and calcined at the required temperature for crystallization. The powdery material is subjected to a milling operation for obtaining fine powders of the desired size and size distribution.

(iii) The wet gel, after drying at 100°C, is milled in alcohol, dried and re-milled to obtain a gel powder. The amorphous gel powder is calcined at the desired temperature to obtain a crystalline oxide powder.

The Pechini method [159], which (with small differences) is also known as the citrate gel method, has been used widely for the synthesis of titanate and other powders. In case of, say, $SrTiO_3$, the process consists in brief of the following steps [160] :

(i) Mixing of a titanium alkoxide with ethylene glycol and citric acid at up to 110°C.

(ii) Addition of freshly prepared strontium carbonate with water and nitric acid.
(iii) Drying at 150°C to obtain a glassy resin, also called a "citrate gel".
(iv) Charring of the resin, crushing of the product and calcination. Recent methods often do without the use of ethylene glycol.

3.3.5 Hydrolysis of Metal-Organics

The techniques of powder preparation by the hydrolysis of metal-organics involve basically the same chemistry as that valid for gel formation in the sol-gel process (Section 3.3.4). In both cases, metal alkoxides or substituted alkoxides have been the most popular choice. The discussion below is therefore based on reactions with alkoxides as starting materials.

Hydrolysis of metal alkoxides, with the general formula $M(OR)_n$ (n = valence of metal M; R = an organic group), is an well-known process for the preparation of single or mixed oxides of high purity and chemical homogeneity, and submicron particle size, going down to nanometric levels [161]. In general, the alkoxides are highly moisture sensitive, and many of them yield fine particles of hydroxides etc. in an uncontrolled way by rapid hydrolysis and condensation as soon as they come in contact with water. Such powders are relatively pure, but are generally composed of heavily agglomerated nanoparticles. The general procedure of such uncontrolled powder preparation is not much different from the sol-gel procedure : the alkoxide is dissolved in an organic solvent (e.g. alcohols, benzene etc.), and the solution reacted with water or an aquo-organic solution. For multicomponent oxide products, two or more alkoxides are used. Typical examples are provided by Hartel and Berglund [162] for TiO_2, Mazdiyasni et al. [163] for $BaTiO_3$, Hardy et al. [164] for $SrTiO_3$ and Ogihara et al. [165] for PZT.

It has been clearly indicated by the above authors [e.g. 162, 164] that the powders prepared simply by the contact between the alkoxides and water were highly agglomerated. Hardy et al. [164] therefore indicated the necessity for controlling the reaction for obtaining powders with controlled characteristics. The seminal work in this area came from Stöber et al. [69], who demonstrated the synthesis of spherical, unagglomerated and almost monosized powders of amorphous silica by reacting silicon alkoxides with water under highly basic conditions.

3.3.6 The Emulsion Process

An emulsion is a heterogeneous system, composed of two immiscible liquid phases, e.g. oil and water, one of which is dispersed under agitation as small droplets into the other [166]. Depending on the nature of the dispersed phase, two types of emulsion result :

(i) water-in-oil type, or W/O emulsion when the dispersed phase is water; and

(ii) oil-in-water type, or O/W emulsion when the oil phase is dispersed in water.

In practical situations, the term "oil" can mean any liquid which is immiscible in an aqueous medium. The dispersed droplets may be of macroscopic or colloidal size. Such emulsions are thermodynamically unstable, as they separate into two distinct liquid layers on standing by the re-coalescing of the dispersed droplets. The instability is caused by the high interfacial tension at the interface of the two liquids. A reduction of the interfacial tension by the use of a suitable emulsifier, i.e. surface active agent gives rise to stable emulsions by preventing the re-coalescence of the dispersed droplets. The droplet size and size distribution commonly depend on the intensity of agitation during the formation of an emulsion.

In a system composed of two immiscible liquids, the dispersion of one phase in the other as small droplets under agitation causes an increase in the surface area of the dispersed phase [166]. Thermodynamically, the increase in the surface area A of the dispersed phase associated with a free energy change can be represented as

$$\Delta G = \gamma \Delta A \qquad \text{... (3.26)}$$

where γ is the surface tension.

The above relationship indicates that for higher interfacial tension, excess energy will be required for breaking up large droplets into smaller one, i.e. for increasing the area of the dispersed phase. The external pressure gradient required for breaking up a droplet of radius a is represented as

$$\Delta P = \frac{2\gamma}{a} \qquad \text{... (3.27)}$$

where ΔP is the Laplace pressure. The equation indicates that a low interfacial tension favours droplet deformation followed by disruption. This can be made possible by the addition of a suitable "emulsifier", which is a (commonly non-ionic) surface active agent.

After the emulsifier has been adsorbed on the dispersed droplets, coalescence of the droplets is generally prevented by the steric hindrance occurring between two droplets. The droplet size and size distribution vary with the amount of the emulsifier present. Thus, the role of the emulsifier during the formation of an emulsion is twofold :

(i) it lowers the interfacial tension between two immiscible liquids, causing thereby droplet deformation and disruption; and

(ii) it prevents re-coalescence of the droplets by adsorbing onto their surfaces, thus producing a stable emulsion.

The above process of obtaining a mixture of two liquids as a dispersed phase and a support phase can be profitably utilized for obtaining solid particles. The simplest steps would be :

(i) Preparation of an aqueous solution of a metal salt.

(ii) Dispersion of this "water" phase in an "oil" phase (e.g. benzene, toluene, hexane, heptane, 1, 1, 1 trichloroethane) under agitation and with the help of an emulsifier.

(iii) Removal of the water or increase in the pH of the solution phase so as to force the precipitation of a solid in the form of particles (salts, hydroxides).

(iv) Washing, drying and calcination of the particles.

When a sol is used instead of a solution, the smooth aqueous sol droplets can be converted to highly spherical gel particles by causing the sol-gel conversion, e.g. by increasing the pH. This can be effected by passing gaseous ammonia ("external gelation") or providing a gelling agent within the system (e.g. urea, decomposing at an elevated temperature) which causes the gelling ("internal gelation"). The spherical gel particles are washed, dried and calcined to obtain highly shrunk oxide spheres. Figure 3.8 shows chromium oxide spheres obtained by an emulsion process [167].

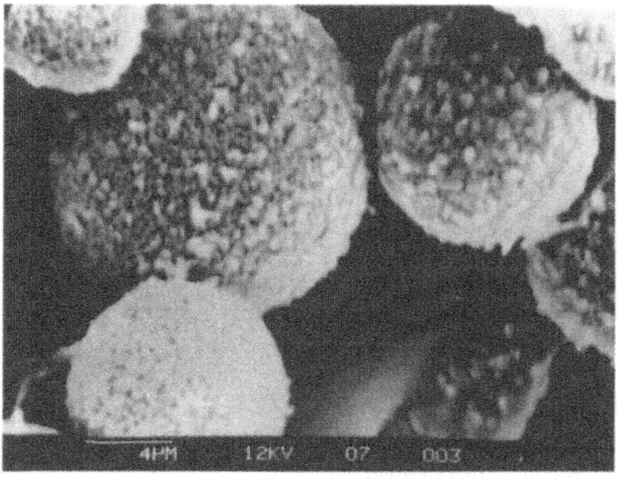

Figure 3.8 Scanning electron micrograph of spherical Cr_2O_3 powder obtained by the sol-emulsion-gel method [167].

The dual nature (also called the amphipathic behaviour) of a surfactant molecule, i.e. a polar "head" and a non-polar "tail" has already been mentioned in Chapter 1. Depending on the charge on the elongated part of the molecule, they are classified into four different groups [23, 168] : (i) anionic (ii) cationic (iii) non-ionic and (iv) amphoteric.

In the anionic surfactants, e.g. sodium stearate, $C_{17}H_{35}COONa^+$, the longer part of the molecule carries a negative charge. Carboxylates, phosphates, sulphonates, sulphates etc. are the characteristic hydrophilic solubilizing groups. In the cationic surfactants, the longer part contains a positive charge, as in cetyl trimethyl ammonium bromide, $C_{16}H_{33}N^+(CH_3)_3Br^-$. The non-ionic surfactants, as the name implies, do not ionize when dissolved in a liquid. The hydrophilic character in this type of surfactants arises from the hydrogen bonding with water molecules.

The non-ionic surfactants used for the stabilization of an emulsion (emulsifying agents or emulsifiers) are of two types : water-soluble or oil-soluble. The most widely used non-ionic water soluble surfactants include the fatty acid esters of anhydro-sorbitols which are made soluble by etherifying the free hydroxyl groups with ethylene oxide (the "Tween" series). When they are not etherified, they thus become good oil-soluble emulsifying agents (the "Span" series) for W/O emulsions. Lauric, palmitic, oleic and stearic acids are used in preparing a series of this type of materials. They are used as emulsifiers, as well as wetting and solubilizing agents. The non-ionic surfactants are generally pH-independent except the esters which hydrolyze in alkaline pH.

The surface activity of non-ionic surfactants is determined by their "hydrophilic-lipophilic (or hydrophobic) balance" or HLB numbers. A high HLB value indicates a strongly hydrophilic character, while a low value is an indication of a strongly hydrophobic nature. Thus, non-ionic surfactants with HLB values in the range (3–6), (8–15) and (15–18) are used as W/O emulsifier, O/W emulsifier and solubilizer respectively [23, 168].

An amphoteric surfactant contains both cationic and anionic groups in the same molecule. They resemble both anionic and cationic surfactants.

Surfactants become effective as emulsifiers when added in small concentrations, e.g. 0.25–5%. Above a critical concentration, the surfactants tend to form oriented molecular aggregates known as "micelles". An excess of emulsifiers can lead also to the formation of the so-called microemulsions, which (like micelles) are thermodynamically stable [169]. Microemulsions are suitable for the synthesis of nanoparticles [e.g. 15] due to the much smaller droplet size than in ordinary emulsions, and still smaller solid particles obtained from them (due to shrinkage during calcination of, specially, gels). The droplet size in emulsions is also substantially reduced by ultrasonication of the water-oil mixture.

3.4 SOLVENT VAPORIZATION

3.4.1 Simple Evaporation

The easiest method of powder preparation from a solution is its evaporation to obtain a powdery solid. The method therefore involves the heating of, say, a soluble salt solution containing one or more components in a suitable container until complete evaporation of the solvent (e.g. water) molecules occurs. The result is the formation of a "dry" solid material. Due to the lack of control on the powder formation process, such products are generally highly agglomerated small particles or an assemblage of coarse particles. As the basic idea is to allow the rapid formation of many nucleation centres and slow growth so that fine particles can be obtained, the solution is generally agitated very rapidly during heating. Another point is the prevention of possible segregation of the different components in the solution during the journey (under heating) to the stage of supersaturation. The dried product is calcined at a suitable temperature (500°C or higher) to obtain oxide powders.

Preparation of lead lanthanum zirconate titanate (PLZT) may be cited as an example of the method [170]. A starting solution of PLZT was prepared by dissolving calculated quantities of $Pb(NO_3)_3$, $La(NO_3)_3.6H_2O$, $ZrO(NO_3)_2$ and $Ti(OC_3H_7)_4$ in acidified water. The clear solution thus obtained became milky during evaporation under heating and stirring. The reactions included the formation of insoluble TiO_2 and the decomposition of the nitrates. The dried multiphase powder was calcined to obtain the desired PLZT powder.

Due to the inherent inadequacies of the process with respect to control on the particle size and size distribution, it is not convenient for drying solutions of large volume. Breakdown of the solution into droplets and subsequent drying is therefore the preferred industrial technique. The major variants of such techniques will be described in the following Sections.

3.4.2 Spray Drying

Spray drying is an old and well-known industrial process for (i) breakdown of a fluid feedstock (solution, slurry, emulsion, sol) into droplets by spraying (also called "atomization") and (ii) drying the sprayed droplets into spherical powder agglomerates during their rapid journey through a column of heated air. The liquid part of the feedstock is commonly water, but can also be aquo-organic or organic in nature. The dried powder is collected in a receptacle. Figure 3.9 shows the basic functions in a laboratory spray dryer.

When the starting material is a ceramic slurry or suspension, the product consists of dried agglomerates of the dispersed ceramic particles in the feedstock. With solutions or sols as the starting material, the obtained spherical agglomerates are composed of salts, hydroxides etc. which are calcined to obtain the corresponding oxides. Agglomerates obtained from both slurries and solutions or sols may need sintering operations to increase the green strength and density or to control the crystalline phases.

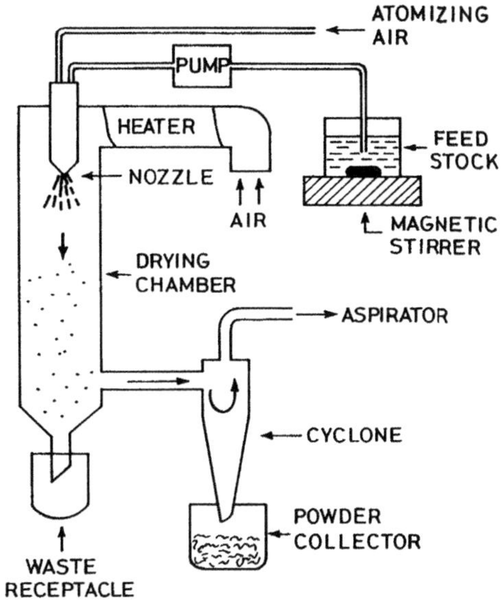

Figure 3.9 Schematic representation of a laboratory spray dryer.

Due to the versatility of spray dryers and their wide applications (they started being used in ceramic manufacturing processes about forty years ago, but were in use in the food or chemical industries even earlier) they come in a variety of configurations and sizes, and new developments are being reported almost every year [171]. An industrial model spray dryer is shown Fig. 3.10. The principles of operation and the feedstock characteristics, described below in brief, are discussed in more detail by Lukaciewicz [172].

Most of the spray dryers use one of the three working principles for atomizing the fluid feedstock :

(a) Rotary (or centrifugal) atomization, where a wheel or disk rotates at the top of the drying chamber at a relatively high speed. The feed which is pumped at a low pressure onto the atomizer, is broken into droplets inside the drying chamber by a centrifugal motion. Different types of feed material can be introduced without using any high pressure pump. The droplet size in rotary atomization decreases with the rotational speed and diameter of the wheel, while the slurry feed rate, viscosity and surface tension of the feed material help to increase the size of the droplet. The particle size can be controlled by the speed of the wheel and also depends on the diameter of the drying chamber. The design is suitable for high powder output requirements.

(b) Pressure nozzle (or hydraulic) atomization, where the feed is atomized by passing it through a nozzle system under a pressure difference; this operation requires a high pressure pump. The droplet size increases proportionately with the size of the nozzle, viscosity and surface tension of the feed material and decreases inversely with the atomizing pressure. The particle size and distribution are controlled mainly by the pressure difference across the nozzle.

Figure 3.10 An industrial model spray dryer (kind courtesy : Dorst-Maschinen und Anlagenbau, Kochel a. See, Germany).

(c) Pneumatic (or two-fluid) atomization where the feed interacts with air at a high velocity. Impingement of air on the feed causes its atomization inside the drying chamber. A feed slurry of almost any desired viscosity can be used. An increase in both viscosity and surface tension increases the average droplet size, while an increase in mass ratio of atomizing air to feed material decreases it. The system works under compressed air and requires a relatively small drying chamber. It is the most suitable design for laboratory scale operations.

Whatever the mode of spray drying, the basic steps in the process remain the same, namely,

(i) Preparation of the feedstock.

(ii) Atomization of the feedstock.

(iii) Drying of the droplets into solid agglomerates.

(iv) Calcination and if necessary, sintering of the products.

A large part of the application of spray drying in ceramics involves the use of an aqueous slurry as the feedstock. The influence of the nature of the slurry on the spray drying process and the quality of the product is known to be significant. Some of the important factors need be pointed out, which has been done below.

A slurry, in its overall characters, resembles a colloidal sol (Section 3.3.4) in that it consists of suspended solid particles in a liquid, but the particle size in the former case is usually much larger, i.e. several microns. In spite of this, the same concept of a balance between van der Waals attractive forces and electrostatic repulsive forces is also valid for predicting the stability (a deflocculated or unagglomerated condition of the particles) of a slurry.

For efficient and economic operation, slurries for sprary drying should satisfy the following conditions :

(i) Proper dispersion of agglomerate-free particles. This requires a pre-milling of the agglomerates and the addition of suitable dispersants.

(ii) Optimally high solid contents. It has been shown, for example, that an increase in the weight % of solids from 50 to 75 can increase the output from the spray dryer by a factor of three [172]. On the other hand, a decrease in the mean particle size can bring down the maximum solid content by close interactions, causing agglomeration and flocculation.

A particle size limit is therefore to be considered while planning a high output. High solid contents, however, lead to choking of the nozzles of spray dryers or other operational problems, and can also lead to inhomogeneities in the slurry. In some cases, the yield does not increase regularly with the solid contents, but exhibits a maximum [173].

(iii) Optimization of the viscosity and density of the slurry. Efficient dispersion and deflocculation ensures a low viscosity even with high solid contents. This avoids clogging of the dryer nozzles during operation, unless the slurry exhibits an increase in viscosity (dilatancy) under shear while being atomized.

The density is controlled not only by the solid content, but also the amount of occluded air (foam content). Foam reduces the density of the spray-dried product and may cause handling losses by creating unwanted and unregulated voids.

(iv) Strength of the product. If the spray-dried agglomerates are required for compaction and sintering of ceramic bodies, it should be ensured that they are "handleable", i.e. they do not generate dust during use. For this, the slurry needs addition of a binder in some small quantity. The binder is usually an organic compound like polyvinyl alcohol, polyethylene glycol etc.

Drying of the droplets is an important factor as it is related to the yield of the material. The drying takes place by the mixing of the droplets with hot air. Depending on the relative directions of flow of the atomized droplets and air, the droplet-air mixing can occur in three ways; the possibilities are described below.

In the co-current method, both the droplets and the drying air flow are in the same downward direction, as the drying air inlet and the atomizer are situated side by side at the top of the chamber, i.e. in a rotary atomizer. In the countercurrent method, the drying air inlet is situated at the bottom, while the atomizer is at the top of the chamber. The third possibility is the mixed flow method, in which the drying air inlet is situated at the top, while the atomizer nozzle is placed at the bottom. Such differences in design have apparently no drastic effects on droplet drying. The drying rate is the maximum just at the time of contact of the droplet and the drying air, which is maintained until the moisture content reaches a critical level, below which the rate falls.

During the spray drying operation, low bulk density powders are often obtained due to the predominance of hollow particles. The major reasons are [172] :

(i) Formation of a film around the droplet, resulting in a reduced surface evaporation rate. The internal liquid vaporizes and expands, causing a ballooning effect.
(ii) Creation of voids in droplets of salt solutions due to the relatively high rate of surface evaporation and salt crystallization.
(iii) Movement of particles from the centre to the surface of slurry droplets, causing internal hollowness.
(iv) Presence of air bubbles in the fluid feedstock.

3.4.3 Spray Pyrolysis

As is evident from the names, spray drying differs from spray pyrolysis in that the former provides for only drying of the solid or hollow spheres obtained as a result of spraying, while in the latter, the atomized (solid) precursor can also undergo heating to a temperature required for its decomposition and conversion to the desired ceramic powder product. Figure 3.11 is a schematic diagram of an early design, showing the basic features.

Many different terms, e.g. “spray roasting”, “evaporative decomposition of solutions”, “solution aerosol thermolysis” and “aerosol decomposition” have been used in the literature to describe basically the same process of spray pyrolysis. The subject has been reviewed in detail by Messing et al. [174]. Briefly, the process consists of the following steps.

(a) Atomization (spraying) of the liquid-rich starting material (mostly true solutions, but also colloidal dispersions, emulsions and sols).

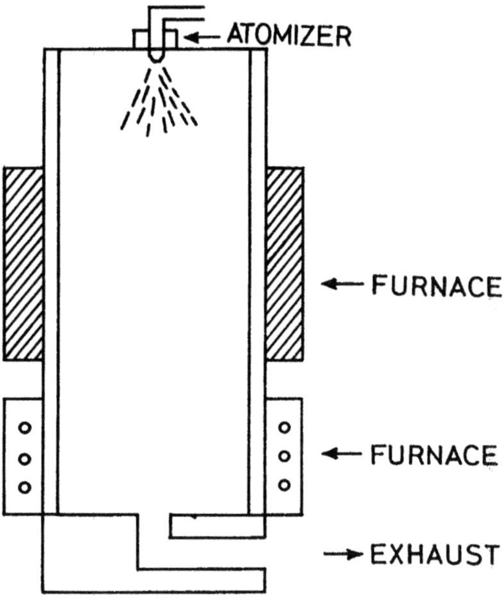

Figure 3.11 Schematic representation of a simple spray pyrolysis system.

Various types of atomizer are in use : pressure type, nebulizers, two-fluid atomizers, and ultrasonic and electrostatic types with varying characteristics. Depending on the density, viscosity and surface tension of the substance to be sprayed, the nebulizer and electrostatic types can generate very small droplets (0.1–10 μm), while the pressure and ultrasonic types produce up to 100 μm droplets. Another important factor responsible for producing small droplets is the frequency of vibration of the atomizing system. The pressure atomizers are apparently more suitable for commercial purposes [174].

(b) Evaporation of the liquid component, solute condensation within the droplet, and drying of the droplets converted into spherical solid particles composed of the solute.

The phenomena to be considered at the beginning are evaporation of the solvent from the surface of the droplet, diffusion of the vaporized solvent going out of the droplet, shrinkage of the droplet and change in its temperature and finally, diffusion of the precipitated solute towards the centre of the droplet. These steps can overlap in the time scale.

Diffusion of the solvent vapour is known to take much less time than that required for the droplet shrinkage. The rate of solute diffusion is considered to be the slowest process at high humidities, but may match with the rate of the droplet shrinkage at low humidities [174]. Another factor is the coagulation of droplets as a result of

Brownian motion before evaporation of the liquid. Obviously the rate of collision or coagulation is proportional to the number of droplets per unit volume which can be controlled. Too low a concentration of the droplets may, however, affect the production rate of the powder.

(c) Pyrolysis (or thermolysis, meaning thermal decomposition) and sintering of the dried precursor spheres.

For obtaining the maximum advantage out of spray pyrolysis, the pyrolyzed droplets should also be made to sinter to high (if not full) density within the reactor. In fact, both the initial laboratory [175] and industrial [176] designs took care of this factor, providing a heating step (up to 1250°C).

It is, however, also understood that the droplet residence time in the reactor is generally less than sufficient for the completion of the pyrolysis and sintering steps. Therefore, to obtain a better control on the latter steps, they are often conducted in separate reactors.

Like in spray drying, the products of spray pyrolysis are often obtained as hollow granules. This has been explained [174] to be due to the lower concentration of the solute at the centre of the droplet than the equilibrium saturation of the solute at the surface during drying. In such cases, surface precipitation, i.e. formation of hollow granules takes place. When the concentration at the centre is the same as or more than the equilibrium concentration at the surface, volume precipitation, i.e. solid granules results. However, if the solute concentration in the droplet is low, the dried granules may be highly porous and tend to be hollow.

The technique is known to be exploited for the synthesis of both single and multi-component powders. Ruthner's spray roasting technology has been used for the commercial production of ferrites and other powders. The method, in principle, has also been used for non-oxide powders [177].

3.4.4 Freeze Drying

All the techniques of removal of the solvent from the precipitated powder, as described in this Section, use heat as the agent.

Freeze drying is a different concept [178] which involves :

(i) introduction of an atomized solution (generally by spraying) into a chamber kept below the freezing point of the solvent;

(ii) freezing of the solvent so as to force the solute to precipitate out; and

(iii) sublimation of the frozen solvent under vacuum.

Figure 3.12 explains the phase relationships involved in the process. The solid lines constitute a schematic of the well-known pressure-temperature equilibrium diagram for water. T indicates the (invariant) triple point where the three phases ice, water and water vapour co-exist in equilibrium. When a salt is added to the system, the freezing point is lowered; a new invariant point P and four phases, i.e. dry salt, saturated salt solution, ice and water vapour are to be considered.

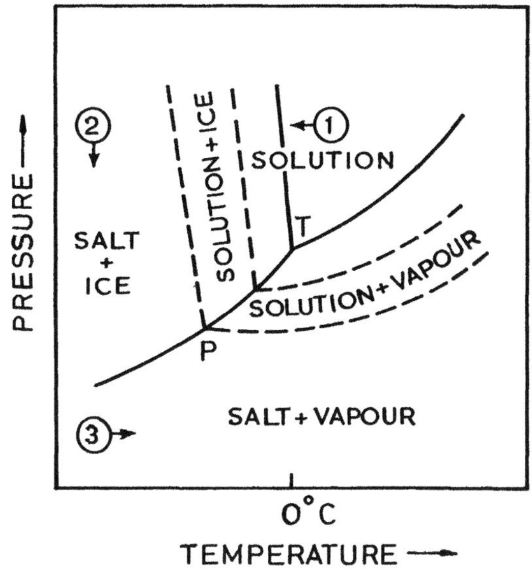

Figure 3.12 Pressure-temperature relationships in the freeze drying process.

When the salt solution at 1 (Fig. 3.12) is suddenly cooled to 2, the water as solvent is converted to ice, so the salt is separated. With a reduction of pressure to 3, i.e. below that of the new invariant point P, the ice evaporates, leaving behind dry powders of the starting salt. Heating decomposes the salt powder to oxide powder.

The simplest case of freeze drying, therefore, consists of the following steps :

(i) The solution is rapidly sprayed into a liquid of low freezing point, much below that of water (e.g. hexane, – 80°C). The droplets freeze to ice; the salt, stripped of the solvating molecules, crystallizes.

(ii) The frozen droplets are treated in a cooled and evacuated condition. The ice is removed by sublimation without melting. Salt particles in soft agglomerates resembling the original spherical frozen droplets are produced. The pioneering work of Schnettler et al. [178] used this technique to obtain aluminium sulphate, magnesium sulphate and magnesium aluminium sulphate to prepare, after calcination at suitable temperatures, Al_2O_3, MgO and $MgAl_2O_4$ spinel respectively. Since then, the freeze drying process has been utilized in the preparation of various other powders [179].

3.5 VAPOUR-PHASE TECHNIQUES

3.5.1 Vaporization-Condensation

Preparation of ultrafine powders and their deposition on suitable substrates by the condensation and deposition of vapours is a relatively old technique. Broadly, vacuum deposition of various thin films and coatings should come under this category [180]. On the other hand, the method has also been used for the preparation of nanoparticles, as described with experimental details of the basic systems by Ichinose et al. [181]. In most cases, the solid charge which is resistance-heated in an evacuated chamber melts before vaporization.

While the above points are generally valid for low-melting compounds and elements (e.g. metals and alloys), ceramic substances because of their refractoriness are generally not easily amenable to such conversion to vapour and controlled condensation to and deposition of powders [182]. Sigel et al. [183] have described a process in which titanium particles are evaporated from a heated tungsten boat at ~ 1550°–1600°C under helium atmosphere to obtain titanium nanoparticles deposited on a cold-finger of liquid nitrogen. The metal particles are then oxidized to TiO_2 nanoparticles by introducing oxygen in the reaction chamber.

3.5.2 Vapour-Vapour Reaction

The vapour-vapour reactions are assuming progressively greater importance in the area of controlled synthesis of ceramic particles, though some reactions are in use for the production of ceramic/glassy powders since long. The reason behind this importance is that extremely fine, nanosized particles, often spherical, and with a relatively low degree of agglomeration can be obtained via homogeneous nucleation of the product(s) of reaction of two or more gaseous components. In addition, the reaction temperature (e.g. in case of AlN) can be much lower than that for competitive processes.

The reaction system generally consists of different variants of the following major parts :

(i) Reactant chamber(s) containing the liquid (to be vaporized) or vapour reactant(s), e.g. $SiCl_4$, SiH_4, alkylaluminiums.

(ii) A carrier gas system for carrying the vapour into the reactor and the products out of the reactor.

(iii) A reaction chamber which is commonly made of 1 metre long impervious tube (specially in laboratory investigations) of alumina, mullite etc.

(iv) A reactant gas (N_2, NH_3 etc.) inlet system.
(v) A heating system (see below).
(vi) A powder collector system.
(vii) A scrubber system for the external handling of corrosive products like HCl.

The heating system can be selected from a variety of options which can be broadly classified as : (a) Furnace heating, (b) Flame heating, (c) Plasma heating and (d) Laser heating.

Heating in a resistance furnace is a classical and most common means of carrying out vapour phase reactions at elevated temperatures. The gases are allowed to react inside the furnace tube, which must be impervious in nature.

The gas flame reactions are also relatively old and the processes are under commercial exploitation for many years. In such reactions, hydrogen-oxygen mixtures are commonly burnt to obtain a flame. Reactants like the tetrachlorides of Si and Ti in a vaporized state are used as the starting materials in flame heating in a commercial scale. The chlorides are oxidized in the gas flame to generate the corresponding oxides, e.g.

$$TiCl_4\ (g) + 2H_2O\ (g) \rightarrow TiO_2\ (s) + 4HCl\ (g) \qquad ...\ (3.28)$$

The products are used in large scale for chemical, pigment and other applications.

In the plasma heating system, the ionized gas produces ions and electrons, which are activated and accelerated by the electric field; they (especially the electrons) play a significant role in determining the average kinetic energy of the gas. This makes it possible to obtain very high temperatures in plasma reactors. The plasma heating system for ceramic powder preparation can be of several types : (i) DC arc plasma, (ii) RF induction plasma, (iii) Hybrid DC-RF plasma.

A commonly used laser heat source is a CO_2 laser operating at 10.591 μm, but with the capability of emitting at different wavelengths between ~ 9.2 and 11.0 μm. In this extensively used system [184–186], a horizontal CO_2 laser beam intercepts a vertical vapour stream of the reactants; the product is collected at the top of the reactor. A simplified sketch is shown in Fig. 3.13. The variability of laser wavelengths gives the system a versatility of matching the frequencies of vibration of the reactant molecules. Various vapour-vapour reactions are made to happen by laser heating; some precursor chemicals and reactions of direct interest are [186] :

$$3SiH_4\ (g) + 4NH_3\ (g) \rightarrow Si_3N_4\ (s) + 12H_2\ (g) \qquad ...\ (3.29)$$

$$2SiH_4\ (g) + C_2H_4\ (g) \rightarrow 2SiC\ (s) + 6H_2\ (g) \qquad ...\ (3.30)$$

$$SiH_4\ (g) + CH_4\ (g) \rightarrow SiC + 4H_2\ (g) \qquad ...\ (3.31)$$

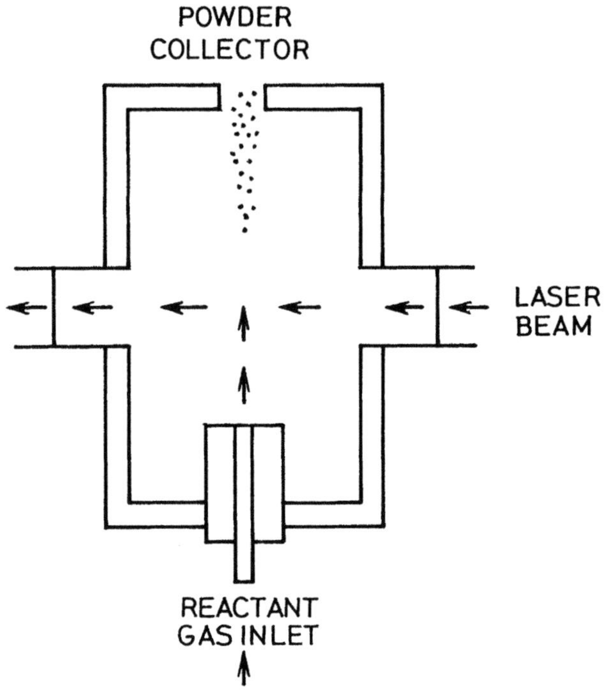

Figure 3.13 Schematic representation of powder preparation by laser heating of vapour reactants.

3.5.3 Vapour-Liquid Reaction

Relatively little has been described in the literature about this kind of reaction, though its potentiality cannot be ignored. It generally leads to precursor powders which have to be thermally treated for conversion to the desired product by decomposition and/or crystallization. One of the specific but important examples is the synthesis of silicon diimide [187] :

$$SiCl_4\,(l) + 6NH_3\,(g) \xrightarrow[0°C]{n\text{-}C_6H_{14}} Si(NH)_2 + 4NH_4Cl \qquad ...\,(3.32)$$

The diimide thus obtained is decomposed at 1200°C to obtain α-Si_3N_4 powder. Note that the modern diimide method for the production of silicon nitride uses a liquid-liquid reaction for obtaining the precursor [188].

A general example, mentioned in Section 3.3.1, is the hydrolysis of metal ions in solution, carried out by the passing of ammonia gas onto the solution surface. Though not used very widely, the technique has the potential of yielding only weakly agglomerated precipitates, and hence, may prove useful in suitable systems. Takase and Sata [189] have described techniques of obtaining alumina particles by the reaction between ammonia gas and $AlCl_3$ solution.

3.5.4 Vapour-Solid Reaction

The vapour-solid reactions (taking into account only the starting materials, not the possible intermediate states) are probably the most well-known and oldest ones — used commercially for a century or more — for the preparation of non-oxide powders. The two major variants are :

(i) Direct synthesis, i.e. nitridation or carbide formation from metal powders.

(ii) Carbothermal reduction of oxide powders for the formation of nitrides or carbides.

Both the reactions commonly take place in a reducing atmosphere with nitrogen or ammonia or their mixtures as the reactant gases.

The direct synthesis is relatively widely used for the production of Si_3N_4 and AlN. In this technique, fine (~ 5 – 20 μm) metal powders, e.g. Si or Al in a loosely packed bed, are allowed to react with nitrogen or nitrogen-ammonia gas mixtures at around 1200°– 1400°C for 10 – 30 h [190]. The overall reactions are :

$$3Si + 2N_2 \rightarrow Si_3N_4 \quad ... (3.33)$$

$$Al + N \rightarrow AlN \quad ... (3.34)$$

$$B + N \rightarrow BN \quad ... (3.35)$$

For obtaining high reaction rates and high quality products several precautions are required to be taken about the quality of the metal powder. The commonly present oxide layer on individual metal particles, unless controlled initially, adds to the oxygen impurity content of the non-oxide powder. Surface nitridation of the particles is another impediment against rapid reaction [191]; in case of AlN, a small fraction of AlN particles is mixed with the reactant aluminium powder [191].

Processes are, of course, available for pure metal particles. Thus, silicon powder is commercially available with a minimum purity of 99.995% and an average particle size of ~ 3.5–7.5 μm [e.g. 192]. Amorphous boron powder is available with an average size of ~ 2–3 μm, but generally the purity is not more than 95%; crystalline powders have better purity [192]. Aluminium powder (atomized or flaky) is also available commercially with the purity level of 99.99%; special powders can be of 99.998% purity [89].

Another point of concern is the state of agglomeration of the metal powder, as this is expected to have a cumulative effect on the agglomeration of the final product. This, in turn, would require intensive milling and cause incorporation of impurities.

In the carbothermal reduction of oxides, the respective oxide powders are mixed with a source of carbon in excess of the stoichiometric requirements to form the starting material. Reactions take place in the range 1100°–1700°C (within about 1400°C for Si_3N_4 and AlN) in nitrogen or other inert atmospheres to yield, e.g. Si_3N_4, SiC, AlN or BN. The carbothermal reduction process is especially important for SiC, as shown by the century-old Acheson process named after the inventor (Section 4.2).

The proposed overall reactions for the carbothermal reduction process are, e.g.

$$3SiO_2 + 6C + 2N_2 \rightarrow Si_3N_4 + 6CO \qquad ...(3.36)$$

$$Al_2O_3 + 3C + N_2 \rightarrow 2AlN + 3CO \qquad ...(3.37)$$

$$SiO_2 + 3C \rightarrow SiC + 2CO \qquad ...(3.38)$$

However, these reactions have some intermediate steps, not discussed here, which involve the reduction of the oxides. In case of SiO_2, the intermediate product is commonly the suboxide SiO; for Al_2O_3, it is the Al metal.

The major drawbacks of the commercially successful carbothermal reduction process, as for SiC, in operation without practically any radical modification since the invention, are :

(i) The excess carbon required for the reaction has to be burnt out at the end, causing the formation of large quantities of CO and CO_2.

(ii) The products are generally obtained in the form of large chunks/ingots which have to be crushed, milled and classified. This allows only limited control on the particle shape and impurity content.

3.6 PRECURSOR DECOMPOSITION

3.6.1 Salt Decomposition

There are many solution techniques of powder preparation which are based on the dissolution and controlled re-crystallization of metal salts. For obtaining oxide particles, the salt particles are then thermally decomposed. The two steps of crystallization and decomposition may be sequential but integrated, as in spray pyrolysis of salt solutions, or separate, as in simple decomposition or the thermal treatment of products of conventional spray drying. In both the cases, the decomposable product is generally a single salt or a mixture of salts, e.g. nitrates. Chapter 4 provides many examples of salt decomposition.

There are also solution techniques which lead to the formation of complex precursor salts, the thermal decomposition of which gives rise to the required multiple oxide product through various intermediate breakdown reactions. One of the most well-known examples is the synthesis of barium titanyl oxalate, $BaTiO(C_2O_4)_2 \cdot 4H_2O$ [193]. Details about its decomposition characteristics have been described in Section 4.1.10.

3.6.2 Polymer Pyrolysis

Inorganic polymers, obtained commonly through the hydrolysis of cations in salts, controlled hydrolysis of metal-organics and the overall sol-gel reactions, have served extensively as precursor substances which, on pyrolysis, have yielded oxide powders.

Polymers like polysilanes and polysilazanes, converting to ceramic compounds (e.g. SiC) on pyrolysis, have been extensively studied. Narula [194] gives a detailed account of such precursors (polymers or monomers); a brief discussion on the topic is also made by Segal [195]. Such polymeric precursors have been shown to be suitable for the preparation of fibres and films, especially the former (e.g. SiC fibres). It has been pointed out by Narula [194] that due mainly to the relatively high cost, these polymeric raw materials may not find application in bulk powder preparation.

Notwithstanding the above, examples are available where the pyrolysis of organic polymers has been resorted to for the preparation of ceramic powders. Thus, Burns and Chandra [196] used polycarbosilanes, polysilanes and polysilazanes for the synthesis of Si_3N_4 powders. The polymers were first heated in graphite or alumina boats in Ar atmosphere for cross-linking at the minimum required temperature (400° – 650°C). The product, ground to 4–8 μm, was pyrolyzed at 400°–1200°C in an atmosphere with 50% ammonia in argon. At 1200°C, amorphous powders of zero or low carbon content were obtained, which were crystallized to α-Si_3N_4 by heating at 1500°C in ultrapure N_2. The polymer precursors were also directly pyrolyzed at 1500°C in 100% electronic grade NH_3. The degree of crystallization varied in the range ~ 40–57 wt%.

Polymer pyrolysis has been extented to various other structures by different workers. Seibold and Rüssel [197] have used polyiminoalane, $\frac{1}{n}[AlH(NR)]_n$ as the precursor for AlN. The precursor was obtained as a gel-like form, which was calcined to 1500°C in ammonia or nitrogen atmosphere to obtain micron-level powders. Interestingly, such powders are believed to be more easily sinterable than those obtained through the conventionl techniques.

4

SYNTHETIC POWDERS : OPTIONS IN PREPARATION

4.0 INTRODUCTION

It has been mentioned elsewhere in this book that there are now numerous options — old and new — available for the synthesis of a given ceramic compound. On the other hand, there are also literally hundreds of ceramic compounds which have been obtained in the form of powder by using one or more (generally the latter) of these preparative techniques.

In this ultimate Chapter, we have selected a few of the more important ceramic materials and described some options of synthesis which have been used in preparing them from different kinds of starting material. Both commercial acceptance and innovativeness have been considered in the choice of the examples. Attempts have also been made to provide examples from the major techniques discussed in Chapter 3. Though the picture is obviously incomplete, it still gives sufficient information on the possibilities in synthesis and their relative importance and feasibility when the preparation of a powder with specific properties (e.g. with respect to size, shape, purity, crystallinity) and/or in economic quantities is planned. It is thus hoped that the recipé-like accounts of synthesis given in this Chapter will help the reader to select specific techniques and build upon them for further innovation.

4.1 SINGLE AND MULTIPLE OXIDE POWDERS

4.1.1 Aluminium Oxide

Corundum or α-Al_2O_3 is thermally the most stable phase among the aluminium oxides. Apart from the α-phase, there are several transition (or "transient") phases of alumina designated by Greek alphabets (Fig. 4.1 shows some of them), which appear at relatively low temperatures during the calcination and conversion of aluminium hydroxides or other amorphous starting materials to α-Al_2O_3. The crystal phases are listed in Table 4.1. The hydroxides and oxyhydroxides that generally form from solutions containing aluminium (like in the Bayer process outlined below) are gibbsite, $Al(OH)_3$, bayerite, $Al(OH)_3$, boehmite, $AlO(OH)$ and diaspore, $AlO(OH)$.

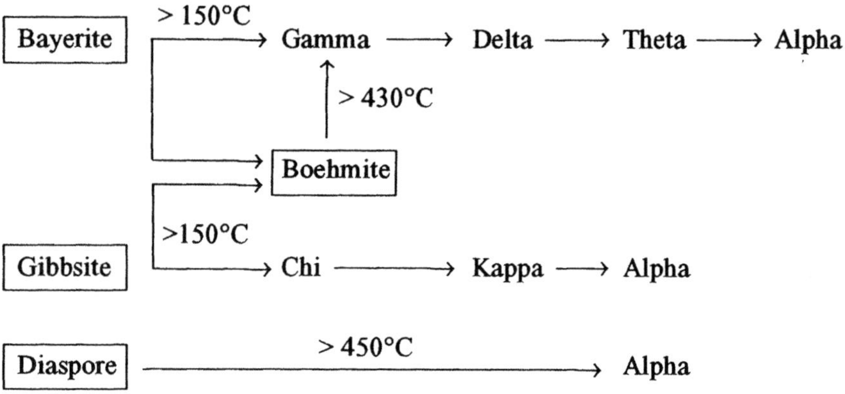

Figure 4.1 Sequence of transformation of hydrated aluminas on heating in air.

Table 4.1 Polymorphs of alumina

Name/Alphabetical Designation	Crystal System
Corundum (α)	Rhombohedral
Gamma (γ)	Cubic
Delta (δ)	Tetragonal
Theta (θ)	Monoclinic
Chi (χ)	Hexagonal
Kappa (κ)	Orthorhombic
Eta (η)	Cubic
Epsilon (ε)	Hexagonal

The Bayer process for the production of alumina which was developed more than 100 years ago by Karl Joseph Bayer [198] uses bauxite ores (Section 2.1.1) as the starting material. Various modifications on the original Bayer process have been made through the last century, but the basic features remain nearly the same. The process, described by various authors [e.g. 90, 199, 200], has the following steps :

(i) Removal of impurities by the physical separation processes.

(ii) Hydrothermal digestion of dry and fine powder of bauxite in a solution of NaOH (and Na_2CO_3 in presence of a flocculant, e.g. starch) at a pressure of 0.5 MPa and ~145°C. Aluminium goes into solution as sodium aluminate, while the impurities SiO_2, TiO_2 and Fe_2O_3 remain as undissolved matter (called "red mud" because of the colour).

(iii) After the digestion is over, the charge is cooled, filtered and seeded with fine particles of gibbsite. Deposition of hydroxides from a supersaturated solution on the seed crystals causes the precipitation (of the "Bayer trihydrate"). The process of precipitation is assisted by temperature, supersaturation of solution, seed content etc. The precipitate is washed and dried. The agglomerated gibbsite particles, which are tens of microns in size, contain the very much undesirable Na_2O as the major impurity, as also SiO_2, Fe_2O_3 etc. in smaller quantities.

During calcination of the product, loss of water starts at 180°C, and the cleaning of pores by this process increases the surface area; the transition aluminas also start to form. The surface area can reach 350 m^2/g at 400°C, above which sintering starts. Alpha alumina starts crystallizing at ~1150°C. The particles retain the morphology of the gibbsite crystals, as also the transition aluminas which precede the formation of α-Al_2O_3. This gives the calcined product a vermicular or "coral" structure.

A lowering of the temperature of formation is generally achieved by the addition of mineralizers like fluorine, chlorine and boric acid; the particle morphology accordingly changes, sometimes drastically (e.g. platy particles with fluorine). Mineralizers that form volatile or water-soluble sodium salts are also used for decreasing the sodium content of the Bayer product. Reduction of the harmful sodium (which, among others, lowers the higher temperature limit of application) by adjustments of the process conditions and washing of the product at appropriate stages has been discussed by Southern [200] and Anderson [90].

The question of soda content has also been addressed recently by de Souza et al. [201] who used Bayer gibbsite as the starting material (vibratory milled to 0.3 μm). A foamy slurry made of the gibbsite, carbon black and polyvinyl alcohol was heated at 1100°–1400°C under a flow of 99.9% argon inside an alumina reactor tube. The processing led to the removal of equivalent Na_2O from 5000 ppm to 10 ppm. Other impurities like Si and Fe were also reduced, though not so dramatically. Further, the process is only exploratory, and requires trials with larger batches.

Various salts of aluminium have been used as starting materials for obtaining alumina via solution routes, but aluminium sulphate, $Al_2(SO_4)_3.xH_2O$ (x = 18 or less) seems to be the most popular. Most of the work involves the precipitation of aluminium hydroxide or a modified sulphate from aqueous sulphate-urea solutions by raising the pH of the system at an elevated temperature through the decomposition of urea (≥ 70°C; Section 3.3.2.). Such reactions, therefore, have been usually carried out at temperatures sufficiently above this, i.e. ~ 90°C. Brace and Matijevic [202] and Fujita et al. [203] were some of the earliest to study the formation of a hydroxide from sulphate-urea solutions via the hydrolysis of a sulphate salt.

Nagai et al. [204] have given a very detailed account of the sulphate hydrolysis reaction, in which a sulphate hydrate was dissolved along with urea in distilled water; the pH was adjusted to 2.0 with HNO_3. The mixed solution was preheated briefly at 50°C before raising the temperature to 90°C under stirring. The reaction proceeded at 90°C without stirring and was stopped when required by cooling the solution to room temperature. The product was thoroughly washed with ethanol and dried at 70°C/3 h under reduced pressure. The agglomerated spherical hydroxide particles decreased in size as a function of (i) increased urea concentration with a constant sulphate concentration, as also (ii) decreased sulphate and constant urea concentration. Agitation of the solution also decreased the particle size, but coalescence of the particles increased with decreasing size.

The hydroxide particles reported by Nagai et al. [204] were about 1 μm in size with a low sulphate concentration (0.005 mol/lit). The product showed a mixed morphology (needle-like pseudoboehmite along with hydroxide spheres) when the reaction was prolonged. Calcination of the dried product at 800°–900°C yielded γ-Al_2O_3 with unchanged particle morphology.

In another work, Nagai et al. [152] observed in addition to the above that prolonged reaction also increased the degree of coalescence of the particles. Following Brace and Matijevic [202], this was explained to be due to the increase in pH of the system with time (guided by the rate of decomposition of urea) towards the point of zero charge (pH ~ 7.0).

In a series of investigations, various authors have synthesized a *basic aluminium sulphate* out of $Al_2(SO_4)_3.18H_2O$ by a similar reaction procedure involving urea [205, 206] with or without NH_4OH/NH_4Cl. Cornilsen and Reed [205], who used a solution pH of 2.3–3.5, found that the basic sulphate precipitate of spherical particles was non-gelatinous, and hence, easily filtered.

An elaborate investigation has been reported by Sacks et al. [206] who used 0.4M and 6.0M solutions of the sulphate and urea respectively in 1 : 1 volume ratio and an equivalent amount of distilled water to prepare the starting solution. Ageing was carried out at 94°C/3h, after which the supernatant solution was immediately separated from the precipitate to avoid re-dissolution. The precipitate was washed by pressure filtration.

A detailed study of the basic sulphate precipitate by Sacks et al. [206] led to the following observations on its character :

(i) The precipitate consisted of agglomerated spherical (1–5 μm) amorphous particles with a specific surface area of about 1 m^2/g.

(ii) The $[Al^{3+}]/[SO_4^{2-}]$ molar ratio was ~ 3.6.

(iii) The sulphate content was ~18 wt% SO_3.

(iv) The amorphous sulphate started showing signs of dehydration at ~130°C which continued to higher temperatures; dehydroxylation was indicated at 430°–510°C.

(v) Crystalline phases like $Al_2(SO_4)_3$, $3Al_2O_3.4SO_3.yH_2O$ (y = 0 – 2) and γ-Al_2O_3 were detected at 600°C. The γ-Al_2O_3 phase increased in quantity and became the only crystalline phase at 900°C.

(vi) Alpha Al_2O_3 started crystallizing at 1050°C and became the only crystalline phase at ~1175°C, though it still contained some residual sulphur. The particle size of this product was < 4 μm and the specific surface area, about 6 m^2/g.

Blendell et al. [207] extended the scope of the synthetic procedure involving Al-sulphate by using formamide in addition to urea and ammonia in the precursor solution. The trends in the rise of pH of a solution containing urea or formamide are reverse—initially fast and later slow for formamide and the opposite for urea. This gave the experimentalist the opportunity to control the pH with time according to the demands on, e.g. the size when both were used. A hydrate phase was produced by ageing near 100°C.

Aluminium sulphate precursor chemicals have also been used for obtaining a calcinable product by freeze drying [208, 209], spray pyrolysis [210, 211] or simply liquid drying [212]. It has been shown [213] that the kinetics of decomposition of the freeze-dried sulphate were faster than the same for a reagent grade sulphate. However, the sinterability of these products might be sensitive (< 40% – 99% of the theoretical density) towards not only the conditions of calcination, but also the residual sulphur content [209]. Crystallization of the freeze-dried products to α-Al_2O_3 required more than 1100°C. The spray-pyrolyzed products, even at 900°C, consisted of a considerable amount of undecomposed salts [210, 211]; these spherical particles, which required > 1000°C to obtain α-Al_2O_3, were hollow and have been recommended for use in porous ceramics or as fillers after calcination. The products gradually decomposed at 500°–850°C, and as expected, yielded a considerable amount of α-Al_2O_3 (with some γ-phase) only at 1100°C.

Okada et al. [210] spray-pyrolyzed five different salts (including Al-sulphate) in usually 0.2 M solutions. Figure 4.2 indicates the structural development of different products of pyrolysis. Note that α-Al_2O_3 was not obtained in any of the cases up to 1000°C. Kato et al. [214] have treated the 18-hydrate with HCl and CH_3COOH and re-precipitated and dehydrated the products with changed $Al^{3+}/(SO_4)^{2-}$ ratios before calcination. At 1200°C/1 h, they crystallized to substantial α-Al_2O_3 mixed with the θ- or η-phase.

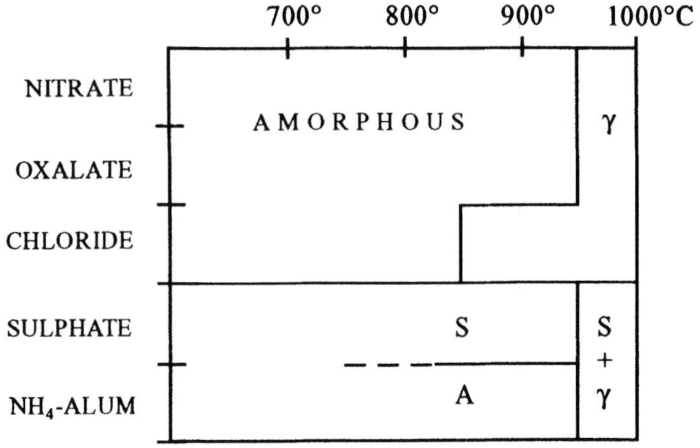

Figure 4.2 Phase developments during calcination of different dried salts of aluminium [210]. S = sulphate, A = alum, γ = transient alumina with spinel structure.

The presence of OH groups in the precursor can sometimes help in lowering the crystallization temperature. One example is the ammonium aluminium carbonate hydroxide, $NH_4AlO(OH)HCO_3$ [215] which was obtained through a reaction between aqueous solutions of NH_4HCO_3 and $NH_4Al(SO_4)_2$, i.e. NH_4-alum. The product obtained at 5°C with a solution pH = 10, when calcined at 1050°C/6 h, yielded 0.2 μm spherical particles of α-Al_2O_3.

Decomposition of high-purity alum crystals is one of the routes for obtaining high-purity alumina particles. The early work on alum decomposition for obtaining alumina particles has been reviewed by Henry and Kelly [216]. The same authors also reported a procedure for obtaining 99.99% Al_2O_3 from ammonium alum. The steps followed were :

(i) Ammonium and aluminium sulphates were dissolved in water at an elevated temperature; the solution was then cooled slowly to allow for the crystallization of the alum. The probable reaction was :

$$(NH_4)_2SO_4 + Al_2(SO_4)_3.18H_2O + 6H_2O \rightarrow (NH_4)_2SO_4.Al_2(SO_4)_3.24H_2O \quad \ldots (4.1)$$

(ii) The large crystals of alum thus obtained were re-dissolved and re-crystallized. Contamination was carefully avoided by discarding all undissolved matter or other particulate impurities by filtration. The large crystals were crushed to ~1.5 cm fragments, apparently avoiding contamination from the crushing system.

(iii) The crushed crystals were dried under vacuum at 180°C, and calcined at up to 1200°C.

The decomposition scheme suggested by Henry and Kelly [216] was :

$$(NH_4)_2SO_4.Al_2(SO_4)_3 \xrightarrow{500^\circ-700^\circ C} Al_2(SO_4)_3 + 2NH_3 + SO_3 + H_2O \quad ... (4.2)$$

$$Al_2(SO_4)_3 \xrightarrow{700^\circ-1100^\circ C} Al_2O_3 + 3SO_3 \quad ... (4.3)$$

The products of calcination were the transient aluminas γ and η (~ 1000°C), transforming to $\delta + \theta$ (~1100°C). Long soaking (23 h) at 1150°C yielded α-Al_2O_3. The average crystallite size for the transient aluminas was 5–10 nm, but increased to > 0.1 μm for α-Al_2O_3 at 1150°–1200°C. A low level of residual sulphur was detected at 1000°C.

The $\gamma - \alpha$ transition starting from ammonium alum was studied in detail by Dynys and Halloran [217]. The decomposition of alum and crystallization of γ-Al_2O_3 (0.05 μm) at 1000°C/2 h was confirmed. As the alum melted at ~ 94°C and the loss of volatiles, i.e. NH_3 and SO_3 took place at 300°–800°C, a foamy calcination product with ~1.5 wt% sulphate was obtained. A ball-milled powder of γ-Al_2O_3 transformed to the α-phase very quickly : > 95 wt% transformation took place after about 3.5 h without an incubation period. Interestingly, the alumina debris from the mill was shown to exert no favourable (seeding) effect on the γ–α transition.

The emulsion technique has been used for the preparation of relatively controlled alumina agglomerates. Thus, evaporation of an emulsion in which the aqueous phase was a salt solution of aluminium has been reported by Sarikaya and Akinc [218]. The emulsion was made of (a) an emulsifier system consisting of a non-ionic surfactant dissolved in a mineral oil at room temperature and (b) an aqueous solution of Al-nitrate added to it under stirring. The emulsion thus produced was added dropwise into a magnetically stirred hot mineral oil bath maintained at 240°C. The emulsion evaporated under a flow of nitrogen, and the particles which precipitated out were removed from the oil by centrifugation. The oil-coated particles were washed thoroughly with toluene, dried and calcined at 1000°C.

The sol-emulsion-gel technique has been shown to produce sufficiently reactive gel spheres to generate α-Al_2O_3 at a relatively short time and low temperature of calcination. Ray et at. [219] who prepared 'alumina' sols from Al-nitrate solutions, have demonstrated this by a calcination study. Ultrasonication is a convenient means of obtaining nanospheroids of alumina from a sol-emulsion-gel process (Fig. 4.3).

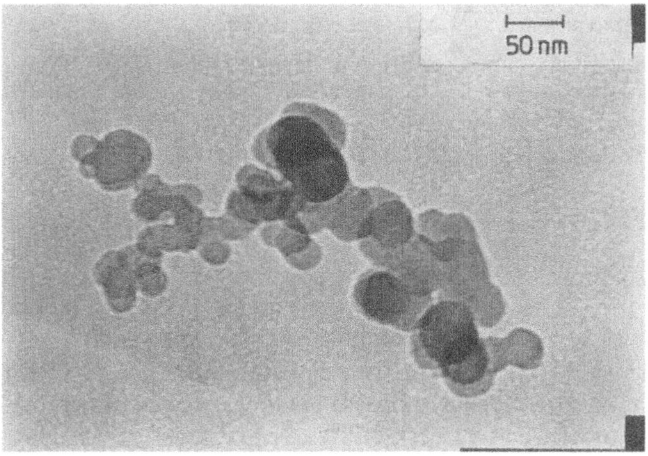

Figure 4.3 Transmission electron micrograph of nanospheroids of α-alumina prepared by the sol-emulsion-gel process under ultrasonication (unpublished work of B. Siladitya, M. Chatterjee and D. Ganguli).

Alkoxide hydrolysis has been used as one of the methods for the synthesis of Al_2O_3 powder, albeit published accounts are relatively few. In an interesting and elaborate work, Ogihara et al. [52] obtained monodisperse, spherical alumina particles through the hydrolysis of Al-alkoxides dissolved in a dilute solution of octanol and acetonitrile. The opposing characters of the two solvents with respect to water afforded a control on the nucleation and growth phenomena. Hydroxypropylcellulose was added as a dispersant. After contact of the reactants, only a few seconds' incubation was required for the initiation of precipitation. The particles, when dried (150°C/24 h) and calcined in air at 400°C, were amorphous. They transformed to γ-Al_2O_3 at 1000°C, and finally, to phase-pure α-Al_2O_3 at 1150°C. Aluminium sec-butoxide was also used by Fanelli and Burlew [220] for hydrolysis, but the product was amorphous and coarse (up to several microns in size). The freeze drying technique has been employed recently for obtaining alumina powders from the isopropoxide [221].

Various vapour phase reactions have been utilized for obtaining fine particles of alumina. The sources ranged from Al metal and alumina to halides like $AlCl_3$. Hirayama [222] used an Al metal (99.999% pure) electrode system in a vacuum chamber provided with a vacuum pumping system and a power source. A mixture of argon and oxygen was introduced in the evacuated chamber. Evaporation of aluminium yielded Al_2O_3 powder carried by the exiting gas to a filter assembly. The product was a mixture of θ- and δ-Al_2O_3 (maximum size ~ 0.1 μm), which was stable against complete thermal transformation to the α-phase (1335°C, compared to ~1280°C in case of a high-purity commercial powder). Again, the specific surface area of this fine powder decreased from 25 m^2/g to 17.3 m^2/g after calcination to 1280°C, while the surface area of the commercial powder dropped drastically from 146 m^2/g to 5.6 m^2/g. This indicated possible efficient use of this powder as a catalyst base material.

Pure alumina was evaporated under neutral and reducing conditions in a plasma furnace by Everest et al. [223] with N_2 as the plasma gas. The rate of evaporation was rather low (6.7 g/min) under neutral atmosphere. It fell significantly with a N_2 – H_2 plasma. However, with addition of carbon and a small quantity of hydrogen as the reducing agents (along with N_2 as the plasma gas), the rate of evaporation shot up to 23 g/min, clearly indicating the choice of atmosphere required for high yield. The final product was spherical γ-Al_2O_3 particles (~ 22 nm) with small quantities of the δ-phase. Flame hydrolysis of $AlCl_3$ is known to produce fumed alumina [224].

Tris(acetylacetonato) aluminium III has been used as the precursor in a vapour phase oxidation reaction [225]. The acetylacetonate was vaporized at 165°C (low supply rate) or 215°C (high supply rate) and mixed with dry N_2 acting as the carrier gas. The mixture was led into a quartz glass/mullite reactor, where it reacted with dry or humid O_2. A complete oxidation was achieved at temperatures greater than 600°C. The particles generated were 10–75 nm in size, depending on the rate of flow of the precursor vapour and temperature. Calcination up to 800°C yielded amorphous powders, which crystallized to γ-Al_2O_3 at 1200°C and δ-Al_2O_3 at 1400°C. Thus, it will be noticed from the above accounts of vapour phase synthesis that α-Al_2O_3 was difficult to obtain through these methods.

The exothermicity of the redox reactions between the decomposition products of metal nitrate(s) and urea during combustion has been utilized by Kingsley and Patil [226] for obtaining α-Al_2O_3. Al-nitrate nonahydrate and urea were mixed (with a small quantity of water) and the paste or solution was heated at only 500°C in a furnace. The heat generated in the reaction increased the temperature to ~ 1600°C, causing the formation of a softly agglomerated foam (Fig. 4.4) of well-crystallized α-Al_2O_3. The agglomerate size was ~ 4.3 μm and the particle size, 0.22 μm. Such reactions, though highly attractive, should be carried out with extreme caution to avoid explosions due to the sudden generation of heat (the entire reaction takes only 5 min).

For obtaining high-purity Al_2O_3, the processes in vogue for many years are alum decomposition, alkoxide hydrolysis and metal oxidation or hydrolysis. All the three kinds of precursor offer the opportunity of high levels of purification. The first two have been already discussed. The metal → oxide procedures are [89] :

(i) Direct oxidation.

(ii) Hydrothermal conversion.

(iii) Reaction with an acid or a base to obtain a hydrated salt or a hydroxide respectively. These methods require calcination of the product to α-Al_2O_3, which in the process releases corrosive acid vapour in the former case and causes cation contamination in the latter.

Figure 4.4 A batch of α-Al_2O_3 powder obtained by the combustion process (kind courtesy : Prof. K. C. Patil, Indian Institute of Science, Bangalore, India).

Martin and Weaver [89] describe an industrial process where Al metal reacts with a quaternary ammonium hydroxide (choline) in a stirred reactor at up to 90°C to produce aluminium hydroxide (gibbsite or bayerite) :

$$\begin{aligned}&(CH_3)_3N(C_2H_4OH)OH + Al + 3H_2O \rightarrow\\&(CH_3)_3N(C_2H_4OH)Al(OH)_4 + 1.5H_2 \rightarrow\\&Al(OH)_3 + (CH_3)_3N(C_2H_4OH)OH \qquad \text{... (4.4)}\end{aligned}$$

It is clear from the above that the process is cyclic, so the reagent consumption is low, and as such, there are no by-products. The obtained trihydrate slurry is spray-dried, calcined and ground. The impurity pick-up of the hydroxide at each step was relatively small, and the final product was more than 99.99% Al_2O_3.

4.1.2 Zirconium Oxide

Zirconium oxide, ZrO_2 exists in three polymorphic forms, i.e. monoclinic (m-), tetragonal (t-) and cubic (c-); only the m-ZrO_2 exists as the thermodynamically stable phase at room temperature. The temperatures of phase transformation in pure ZrO_2 are [227] :

$$\text{Monoclinic} \underset{950^\circ C}{\overset{1150^\circ C}{\rightleftarrows}} \text{tetragonal} \xrightarrow{2370^\circ C} \text{cubic} \xrightarrow{2680^\circ C} \text{melt}$$

The tetragonal to monoclinic transformation, on cooling through the transition temperature, is associated with a considerable volume expansion (3–5%) and a shear strain (1–7%) causing catastrophic cracking in a pure (i.e. undoped) zirconia ceramic. To overcome this difficulty, different amounts of structure-stabilizing dopant cations, e.g. Mg^{2+}, Ca^{2+}, Y^{3+}, Ce^{4+} etc. are incorporated in ZrO_2 [228]. Depending on the amounts of the stabilizing ion, partially stabilized zirconia (PSZ) or fully stabilized zirconia (FSZ) can be formed. In presence of only an insufficient amount of stabilizing ions, the PSZ obtained contains a mixture of c- and t- and/or m-phases; t-ZrO_2 transforms to m-ZrO_2 on cooling [228]. When a sufficient amount of the stabilizer is present, FSZ containing a single phase c-ZrO_2 is formed.

For the preparation of ZrO_2 powders, the production routes of commercially viable processes normally use the economically available natural sources, namely, baddeleyite (ZrO_2) and zircon ($ZrSiO_4$). Of the two, zircon is more important because of its wide availability as beach sand deposits. There are various processes for the extraction of zirconium from zircon. They are classified into four groups, as described below.

(a) Chlorination and thermal decomposition technique. An intimate mixture of zircon and carbon is obtained after milling and pelletizing. When this is directly chlorinated at 800°–1200°C in a shaft furnace or fluidized bed, the following reaction occurs [228, 229] :

$$ZrSiO_4 + 4C + 4Cl_2 \rightarrow ZrCl_4 + SiCl_4 + 4CO \qquad \text{... (4.5)}$$

Zirconium tetrachloride, after being distilled, is selectively condensed at 150°–180°C when the major impurities, i.e. chlorides of silicon, iron, titanium and aluminium are separated. After hydrolysis with water, a solution of zirconium oxychloride or zirconyl chloride ($ZrOCl_2$) is obtained. The $ZrOCl_2$ solution is used for the preparation of ZrO_2 powders.

(b) Alkali oxide decomposition technique. Zircon may be decomposed either with NaOH [230] or Na_2CO_3 [231] which act as the fluxing agents. In presence of NaOH, zircon decomposes at 600°–700°C, forming sodium zirconate, sodium zirconate silicate and sodium silicate. The degree of decomposition depends on parameters like particle size of zircon, alkali/zircon molar ratio, temperature and time of decomposition etc. By careful control of the process parameters, the following reaction products are obtained:

$$ZrO_2.SiO_2 + 4NaOH \rightarrow Na_2ZrO_3 + Na_2SiO_3 + 2H_2O \quad \text{... (4.6)}$$

In presence of Na_2CO_3, the conversion occurs at a much higher temperature, e.g. 1000°–1100°C, as follows :

$$ZrO_2.SiO_2 + Na_2CO_3 \rightarrow Na_2ZrSiO_5 + CO_2 \quad \text{... (4.7)}$$

$$ZrO_2.SiO_2 + 2Na_2CO_3 \rightarrow Na_2ZrO_3 + Na_2SiO_3 + 2CO_2 \quad \text{... (4.8)}$$

Leaching with water of the decomposed mass removes the soluble sodium silicate and simultaneously hydrolyzes sodium zirconate to hydrated zirconia. The impure hydrated zirconia is dissolved in concentrated HCl, HNO_3 or H_2SO_4 at 80°C for obtaining a zirconyl chloride, nitrate or sulphate solution respectively.

These solutions are then purified either by precipitation as basic zirconium sulphate [232–234] or by crystallization as the corresponding zirconium salt [230, 234]. Further chemical treatment of the precipitates or crystals, as described in the subsequent Sections, generates ZrO_2 powders. The Na_2ZrSiO_5 phase is decomposed by the action of mineral acids. Silica precipitates out and is separated; the rest forms an impure zirconyl salt which is purified [234].

(c) Lime fusion technique. When a mixture of $ZrSiO_4$ and calcia (in different proportions) is heated at various temperatures, calcium zirconium silicate, calcium zirconate and calcium silicate are formed [228] :

$$ZrO_2.SiO_2 + CaO \xrightarrow{1100^\circ C} CaZrSiO_5 \quad \text{... (4.9)}$$

$$ZrO_2.SiO_2 + 2CaO \xrightarrow{1600^\circ C} ZrO_2 + CaSiO_3 \quad \text{... (4.10)}$$

The second reaction is more suitable because $CaSiO_3$ can be removed by a treatment with HCl. When doloma (calcined dolomite) is mixed with zircon, followed by heating, various mixtures of zirconia and calcium magnesium silicate are produced. In presence of an equimolar mixture of $ZrSiO_4$ and limestone, heated at 1100°–1500°C/3–9 h, decomposition of zircon occurs as follows [235] :

$$ZrSiO_4 + CaCO_3 \rightarrow ZrO_2 + CaSiO_3 + CO_2 \quad \text{... (4.11)}$$

From the mixture of ZrO_2 and $CaSiO_3$ thus obtained, $CaSiO_3$ is removed first by leaching with HCl at 90°C followed by treatment with NaOH at 15°C.

(d) Plasma decomposition technique. When zircon particles are introduced into a stable argon plasma reactor at a temperature range of 6000°–15000°C and quenched rapidly, dissociation of zircon occurs with the formation of the constituent oxides, i.e. ZrO_2 and SiO_2 [236]. The degree of dissociation depends on a number of process parameters, e.g. the particle size of zircon, flow rate of Ar gas, arc current, feed rate etc. The decomposed mass is treated with NaOH or H_2SO_4.

(e) Zircon can be thermally dissociated in an arc furnace or an electric reactor [229] when heated above 1750°C (followed by rapid cooling), into its constituent oxides. ZrO_2 and SiO_2 are recovered separately as discussed above.

The most simple and commonly used method for the preparation of zirconia powders is the precipitation/co-precipitation of the metal salt solution. Some examples are given below.

Itoh [237] prepared ZrO_2.2.9 mol% Y_2O_3 powder using $ZrOCl_2$ and YCl_3 solutions as the starting materials. The precipitation was conducted with NH_4OH solution at pH = 8. The hydroxide precipitate was first washed with distilled water and subsequently with ethanol. Calcination of the washed precipitate was carried out at 500°–1300°C. Only t-ZrO_2 was obtained at 600°C. Some m-ZrO_2 appeared along with t-ZrO_2 at 800°C. Submicron particles were obtained by this method. To obtain Er_2O_3 doped ZrO_2 powder, Duran and co-workers [238] followed this technique using Er_2O_3 and zirconium tetrabutoxide as the source materials. The water-isopropanol washed precipitate was calcined at 550°C/1h. The calcined powder contained soft agglomerates of small crystallites (~ 8.5 nm); t-ZrO_2 crystallized at this temperature.

Interestingly, addition of SO_4^{2-} ion (through H_2SO_4) [239, 240] into the water-washed hydrous precipitate followed by calcination helped to retain the metastable t-ZrO_2 in the calcined powder. The presence of a moderate amount of SO_4^{2-} in the precipitated material remarkably increased the amount of c/t phases [240]. Again, Hofmann et al. [241] showed that the addition of titanium in the system ZrO_2.3 mol% Y_2O_3 increased the stability of t-ZrO_2. Due to such addition, the hydrolysis of Y_2O_3 in the humid atmosphere was retarded.

Van de Graaf and Burggraaf [242] prepared agglomerate-free ZrO_2 powder following the procedure applied by Murata et al. [150] for PLZT. The method involved the addition of hydrogen peroxide to a mixed aqueous solution of $Y(NO_3)_3$ and $ZrOCl_2$ in the molar ratio of 1 : 1 of the metals. Co-precipitation of Y^{3+} and Zr^{4+} ions as hydroxides at 40°C was carried out by dropwise addition of the salt solution to the ammonia solution under stirring at pH = 8.5–9.5. The ZrO_2–Y_2O_3 powder of high surface area and low agglomerate size was obtained after calcination of the water-washed hydroxide precipitate at 600°C/2 h.

In some cases, ZrO_2 powder has been prepared by the hydrolysis of the metal salt solution at elevated temperatures, e.g. 65°–100°C (forced hydrolysis) [243, 244]. Hydrolysis of a highly acidic solution of $ZrOCl_2.8H_2O$ containing K_2SO_4 and HCl was carried out at 65°–100°C for obtaining hydrated zirconia [243]. Under optimized conditions, monodisperse hydrous zirconia particles of spherical morphology were obtained. The particle size of the hydrous material varied from 0.1 to 2.0 μm depending on the experimental conditions. At 400°C, t-ZrO_2 crystallized, which at 900°C transformed to m-ZrO_2. An average crystallite size of 7.4 nm was obtained in the powder calcined at 550°C.

Ultrafine ceria-stabilized zirconia powders containing 12 mol% CeO_2 and that coated with 2 mol% Y_2O_3 were prepared by dissolving zirconium sulphate tetrahydrate and cerium nitrate hexahydrate in distilled water followed by the addition of polyvinylpyrrolidone and urea [244]. Hydrolysis of the solution at 85°C/3 h produced particles which were washed with water and isopropanol and dried at 100°C. Ceria-stabilized ZrO_2 powder with spherical morphology and with a narrow size distribution of 400–700 nm was obtained after calcination of the dried powder at 500°C/2 h . Coating of the powder with 2 mol% Y_2O_3 was carried out at pH = 2 and at 850°C/1 h using yttrium chloride solution and urea.

Several alternatives exist for the synthesis of ZrO_2 powders (Fig. 4.5) by the hydrothermal technique. Tani et al [245] precipitated amorphous zirconia from Hf-free $ZrCl_4$ solution with NH_4OH, making the precipitate chloride-free by washing with distilled water followed by drying at 120°C/4 h. The hydrous mass

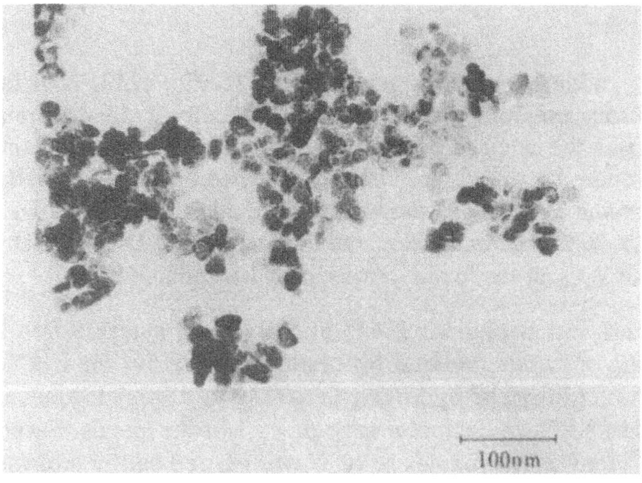

Figure 4.5 Hydrothermally crystallized zirconia nanoparticles
(kind courtesy : Prof. S. Somiya, The Nishi Tokyo University, Tokyo, Japan).

was treated hydrothermally in presence of different types of mineralizer, e.g. KF, NaOH, LiCl, KBr in the temperature range of 200°–600°C and under a pressure of 100 MPa. Depending on the experimental conditions and type of the mineralizer, submicron particles of m- and/or t-ZrO_2 were obtained without further heat-treatment. For the preparation of stabilized ZrO_2, different types of salt of the doping metal, e.g. $MgCl_2$ [246] and YCl_3 [247] have been used for co-precipitation along with $ZrOCl_2$. The co-precipitated material was then treated hydrothermally at 220°–250°C. Submicron particles of c-, m- and t-ZrO_2 were obtained. The reactions involved during the formation of ZrO_2 powder from Zr and $ZrSiO_4$ have been described in Section 3.3.3.

Bleier and Cannon [248] generated ZrO_2 powder by treating zirconyl salt solutions (except zirconyl sulphate, which produced gels) directly under controlled hydrothermal conditions, e.g. 98°C/70 h. Under such conditions, hydrolysis of the zirconyl salts produced a complex species, $[Zr(OH)_2.4H_2O]_4^{8+}$, having less solubility than that of the source materials. A growth of the complex species was believed to produce m-ZrO_2 powder which was porous in nature. The generated particles (~ 80 nm) consisted of 3 nm crystallites. High-purity $ZrOCl_2.8H_2O$ solutions with additives like NaOH, Na_2CO_3, Na_2SO_4, NH_4F, HNO_3 and H_2SO_4 [249] have also been used for the preparation of ultrafine ZrO_2 powders.

Zirconia gels, prepared from zirconium isopropoxide in alcoholic medium, were also treated hydrothermally for conversion to ZrO_2. In a series of experiments, Komarneni et al. [250] obtained t- and m-ZrO_2 powders having the smallest particle size of 5–6 nm and 10–30 nm respectively.

Monoclinic ZrO_2 powders of acicular morphology have been prepared by the hydrothermal treatment (250°C) of a solution obtained by dissolving zirconyl carbonate in H_2SO_4 solution [251]. A metastable intermediate compound was believed to be formed which after treatment for a long time formed acicular m-ZrO_2 particles of two types : (i) flaky particles of oblong shape and (ii) rod-like particles 0.3 to 1.3 μm in length and 0.1 to 0.2 μm in width. Addition of $MgSO_4$ to the system facilitated the crystallization.

Kato et al. [252] generated zirconium oxide sulphate crystals of chemical composition $Zr_3O_5SO_4.nH_2O$ by the hydrolysis of an aqueous solution of $ZrOSO_4$ at 240°C. The zirconium oxide sulphate particles were treated hydrothermally at 240°C in presence of a dilute H_2SO_4 solution. During this treatment, zirconium oxide sulphate decomposed gradually with the formation of H_2SO_4 in increasing quantity. The formation and morphology of m-ZrO_2 were automatically controlled with the generation of H_2SO_4 during decomposition of the sulphate and were influenced by the morphology of the oxide sulphate crystals. Under optimized conditions, anisotropic m-ZrO_2 single crystals were formed.

Spray pyrolysis has been used widely for the preparation of ZrO_2 powders. In a typical process, Dubois et al. [253] prepared ZrO_2.3mol% Y_2O_3 powder by the spray pyrolysis technique, using $ZrOCl_2.8H_2O$ and $Y(NO_3)_3.5H_2O$ as the source materials. A mist of the mixed solution of Zr^{4+} and Y^{3+} ions with a narrow size distribution was generated by an ultrasonic atomizer at different frequencies and introduced into a furnace along with a carrier gas below 400°C so as to avoid crystallization of the particles. The powders collected with an electrofilter were calcined in air at 410°C/1 h and then at 700°C/2 h when the volatiles were totally removed. The specific surface area of the powder was 30 m^2/g with the mean particle size of 0.2–1.8 μm.

In another method, Yoshioka and co-workers [254] prepared t-ZrO_2 of the composition ZrO_2.12 mol% CeO_2 by the spray pyrolysis of a mixed solution of $ZrO(NO_3)_2$ and $Ce(NO_3)_2$. The product at 400°C was amorphous and started to crystallize to t-ZrO_2 at 500°C with a maximum particle size of 0.68 μm; at 700°C, the size of t-ZrO_2 was 0.42 μm. The presence of redox compounds, e.g. NH_4NO_3 and NH_2CONH_2 as additives in the spraying solution helped in the production of fine particles.

During the preparation of pure and Y-ZrO_2 by spray pyrolysis with $ZrOCl_2.8H_2O$ and $Y(NO_3)_3.6H_2O$ as the starting materials, Xiaming et al. [255] found that both the calcination temperature and the flux of the carrier gas controlled the formation of crystalline phases in the powder particles. With an increase in the calcination temperature from 500° to 1000°C, the amount of t-ZrO_2 formed went on decreasing. With an increase in the flux of the carrier gas, the residence time of the particles in the reaction chamber decreased with the formation of less m-ZrO_2. The effect was more pronounced in Y-ZrO_2 even at high temperatures.

Fine ZrO_2-Y_2O_3 powders containing 2–6 mol% Y_2O_3 were also prepared by the spray pyrolysis of mixed zirconium and yttrium alkoxide solutions [105]. For this purpose, an anhydrous ethanolic solution of Zr-n-butoxide containing a calculated quantity of yttrium isopropoxide, was first refluxed in N_2 atmosphere at 78°C/1 h. A mist of the alkoxide solution was first passed into a furnace at 400°C, which was allowed to flow along with N_2 to a second furnace at 700°C. The powders were calcined at 800°– 900°C/1 h. The average particle size, the specific surface area and the average crystallite size of the spherical powders were 0.24 μm, 8.7 m^2/g and 40 nm respectively. The alkoxide-derived powders were solid and free from agglomeration. The sphericity of the powder was dependent on the refluxing time of the solution.

Chatterjee et al. [173] prepared ZrO_2 powders by the spray drying of aqueous slurries of different ZrO_2 contents, e.g. 5–25 g/l. The spray dried materials were calcined at 1000°C/0.5 h. The average particle size of the calcined powders was

3–6 μm. The as-sprayed powders showed specific surface areas of 151–212 m^2/g which were drastically reduced to 5–7.5 m^2/g after calcination at 1000°C/0.5 h. All the calcined samples contained t-ZrO_2 as the only phase.

In another experiment, Chatterjee and Ganguli [256] obtained pure, MgO (7 mol%) and CeO_2 (7 mol% and 25 mol%) doped ZrO_2 powders by spray drying of aqueous solutions of pure zirconyl nitrate and that containing calculated quantities of $Mg(NO_3)_2.6H_2O$ or ammonium cerium (IV) nitrate. Spraying was carried out at 180°C. The as-prepared powders were highly spherical in morphology with diameters ranging from 0.4 to 2.5 μm. The precursor powder was calcined at 700°–1200°C for 2 h. All the calcined samples, both pure and doped, showed the presence of c-ZrO_2 as the only phase at 700°C. At 1000° and 1200°C, only the CeO_2-doped powders contained phase-pure c-ZrO_2.

For the preparation of freeze-dried ZrO_2.10 mol% Y_2O_3 powder, sulphate solutions of Zr^{4+} and Y^{3+} were used [257]. The method involved the spraying of the mixed sulphate solution into liquid nitrogen. After freezing, the mixed sulphate solution formed a low-density sponge-like material. On drying for 15 h in vacuum at 1.3–0.13 Pa, the residual water was removed at 150°C. The oxide material was obtained at 800°–1100°C. Complete conversion of the sulphate material to oxide powder (10–20 nm) occurred at 900°C/17 h.

Hydrolysis of zirconium alkoxide is known to produce high-purity zirconia powder. Mazdiyasni and his group [258] applied this method for the preparation of ZrO_2 powders containing different mol% of Y_2O_3. In a typical process, high-purity zirconium isopropoxide, $Zr(OC_3H_7)_4$ and yttrium isopropoxide, $Y(OC_3H_7)_3$ were dissolved in an organic solvent, e.g. benzene or n-hexane. Hydrolysis of this mixed solution was effected by dropwise addition of triply-distilled water to the vigorously stirred mixed alkoxide solution. Quantitative precipitation of the mixed oxide occurred.

In place of the isopropoxide, other alkoxides, e.g. butoxides have also been used. For the more stable alkoxy group, addition of a small amount of acid accelerated the reaction. The method produced high-purity, chemically homogeneous powders with an average particle size of ~ 5 nm; after calcination at 850°C/0.5 h, the particle size increased to ~ 40 nm.

Hydrolysis of Zr-n-propoxide has been controlled by adding a chelating agent, e.g. acetylacetone (acac) for obtaining zirconia [259]. In such a case, the hydrolysis was carried out by adding water diluted with isopropanol to a solution of the alkoxide and acetylacetone diluted with isopropanol under stirring. Keeping the H_2O/Zr molar ratio fixed at 5, an increase in the molar ratio of acac/Zr from 0 to 0.8 caused a change in the nature of the precursor material from a colloidal precipitate to polymeric gels. Calcination of the powdery precipitate produced stable m-ZrO_2, while the gel material crystallized above 500°C to metastable t-ZrO_2 as the inital phase.

It was shown by Yoldas [260] that the morphology and the particle size of the ZrO_2 powder derived via hydrolysis and condensation of zirconium alkoxides were strongly dependent on the type of the alkyl groups present in the alkoxides, water/alkoxide molar ratio, reaction temperature and medium and the molecular separation of the species. Both the sintering behaviour and the crystalline transformation of t-ZrO_2 to m-ZrO_2 were in turn affected by the morphology and particle size of ZrO_2 obtained from the alkoxide. Addition of SiO_2 increased the stability of t-ZrO_2.

Fegley et al. [261] prepared monodisperse, pure and doped zirconia powders of controlled shape, size and size distribution by the hydrolysis of an ethanolic solution of Zr-n-propoxide or isopropoxide with an excess of water diluted with ethanol at 50°C or 25°C respectively. For the doped powders, Y-isopropoxide was co-hydrolyzed.

In both the cases, unagglomerated, spheroidal powders with an average particle size of ~ 0.2 μm were obtained. Specific surface areas of ZrO_2 and ZrO_2–Y_2O_3 powders were 28–58 m^2/g (after ageing), and 66–77 m^2/g respectively for the water-washed powders. The as-prepared powders were amorphous. After calcination at 600°C/4 h [261], pure ZrO_2 powder contained the m-phase and the YSZ powder showed the presence of c- and t-ZrO_2. During the preparation of 3 mol% Y-doped ZrO_2 powders, Uchiyama et al. [104] found the refluxing of the mixed alkoxide (~75°C) to be essential before hydrolysis.

Lerot et al. [13] synthesized monodisperse, generally spherical, size-controlled zirconia powders by the controlled hydrolysis (25°C) of Zr-n-propoxide in alcohols in presence of long-chain (C_6–C_{18}) carboxylic acids.

A method of preparing monodisperse spherical ZrO_2 powder, developed by Rinn and Schmidt [262], consisted of the following steps. An ethanolic solution of hydroxypropylcellulose (HPC) and different amounts of acetylacetone (acac) were added to an ethanolic solution of Zr-isopropoxide. Addition of a mixture of nitric acid and water followed by heating at 40°C caused precipitation. After treating the suspension with petroleum ether, particles were collected by centrifugation. Spray drying of the suspension, prepared in petroleum ether under ultrasonication, produced monodisperse spherical powder with a significant flexibility in the average size.

Another well-known technique of ZrO_2 powder preparation is the sol-gel method. In the following, some examples of the preparation of such powders from both inorganic salts and alkoxides are discussed.

A ZrO_2.2.7 mol% Y_2O_3 powder [263] containing 5 wt% Al_2O_3 was prepared first by dissolving $ZrOCl_2.8H_2O$ and $AlCl_3.6H_2O$ in water; thereafter, Y_2O_3 dissolved in dilute HCl solution was added to it. A white sol was obtained at pH = 4 when dilute NH_4OH was slowly added to the above solution under stirring. At pH = 6, a white gel was obtained. The gel after drying at 100°C was milled in alcohol. The resulting powder was finally crystallized at 500°C/15 h.

Hydrolysis and condensation reactions involving Zr-n-butoxide in 2-propanol and water with nitric acid under stirring at room temperature have been used to generate ZrO_2 particles [264]. A stable zirconia sol containing monodispersed particles of 4 nm size in water was prepared by removing the solvent, i.e. 2-propanol by evaporation. By adding yttrium nitrate hexahydrate, stable composite sols were obtained; gels were formed after drying the sols. Calcination of the gels at 300°–1100°C/5 h produced oxide powders. Pure zirconia gels produced m-ZrO_2 at 500°C which persisted up to 900°C; on the contrary, c-ZrO_2 was obtained as the single phase from gels containing 8 mol% Y_2O_3 at 500°C which persisted up to 900°C. The crystallite size of the ZrO_2.8 mol% Y_2O_3 at 500°C was 6 nm which was much smaller than that of pure ZrO_2 (11 nm).

Ceria-doped ZrO_2 powder (0.84 ZrO_2.0.16 CeO_2) was synthesized by the hydrolysis of a mixture of Ce (III)-2, 4-pentanedionate and Zr-n-butoxide in anhydrous ethanol with water [265]. After keeping the mixture at a simmering temperature, the excess ethanol was distilled off. A small amount of water was added to the resulting viscous liquid and the semi-solid mass obtained was allowed to dry at 60°C. The gelled material was calcined at 550°C and had a specific surface area of 11 m^2/g. Calcination from 300° to 1050°C generated t-ZrO_2 as the only phase.

Free-flowing powders generally with spherical morphology have been prepared by the sol-emulsion-gel method. The zirconia sol for this type of experiment was produced by peptizing hydrous precipitates of zirconia, obtained from $ZrOCl_2$. $8H_2O$ solution, with nitric acid at 80°–90°C under stirring [266]. A commercially available zirconium hydroxide paste, peptized with HNO_3 [157] or zirconyl acetate solution [267], was also used. Dopant cations, e.g. Y^{3+} ion were added to the Zr^{4+} sol/solution. The gel microspheres thus obtained were filtered, washed with the organic solvent used for emulsion preparation and dried at 100°C. Micron-sized stabilized zirconia spheres were obtained after calcination of the gel spheres at the desired temperature. The gel spheres retained their sphericity even after calcination at high temperatures.

Nanometer-sized ZrO_2–Y_2O_3 powder [268] was obtained following the sol-emulsion-gel technique with ultrasonication. Water was removed from the reaction mixture by distillation. The gel material after drying at 140°C was calcined from 400° to 1350°C. The spherical clusters were found to contain 4–10 nm spherical particles in samples calcined at 600°C and formed soft agglomerates. Metastable t-ZrO_2 was found up to 650°C, while m-ZrO_2 formed above 700°C. At 200°C, the powder showed a specific surface area of 364 m^2/g, which went on decreasing with increase in the calcination temperature; at 600°C, the value was 78 m^2/g.

Thermal decomposition of citrate precursors is also known to produce ZrO_2 powders. Van de Graaf and Burggraaf [242] synthesized Y- and Gd-doped ZrO_2 powders using $ZrOCl_2.8H_2O$ and Y_2O_3 and Gd_2O_3 as the starting materials. A zirconyl nitrate solution containing Y or Gd as the dopant was prepared from $ZrOCl_2.8H_2O$ and HNO_3. The pH of the solution was brought to 6–7.5 with ammonia followed by the addition of citric acid so as to obtain an organo-metallic complex.

The highly viscous material, obtained after partial dehydration, was thermally treated for continuous evaporation of the liquid. At this stage, swelling of the material took place due to the removal of the decomposed gases from NH_4NO_3 present in the system, when ignition occurred followed by continued exothermic reaction. The precursor material thus obtained was calcined at 650°C in air when stabilized ZrO_2 powders of loosely packed tissue-like morphology and with a specific surface area of 58 m^2/g were obtained.

Srivastava and Dongare [269] prepared a glassy precursor material by evaporating a mixture of $ZrOCl_2.8H_2O$, citric acid and water at 35°C. The precursor material produced t-ZrO_2 particles (0.15–0.2 μm) at 400°–500°C. Following the same procedure, ZrO_2.5 mol% CaO was prepared by adding $CaCO_3$ to a $ZrOCl_2.8H_2O$–citric acid mixture. A precursor glassy material was obtained at ~ 80°C by evaporation. Calcination of the precursor material from 300° to 1250°C in O_2 atmosphere produced c-ZrO_2 (6–41 nm) crystallites.

For the preparation of ZrO_2–CeO_2 powders containing 20 mol% and 50 mol% CeO_2, Yashima et al.[270] used $Ce(CH_3COO)_3.H_2O$ and $ZrOCl_2.8H_2O$ as the starting materials. A mixed solution of Zr^{4+} and Ce^{3+}, after dissolution in an anhydrous citric acid and ethylene glycol mixture at 100°C, was further heated under stirring at 140°C until a highly viscous material was obtained. The viscous mass at 350°C produced a black, amorphous precursor powder. At 400°C/1 h, the precursor self-ignited with the formation of a yellow powder ready for pressing and calcination.

Thermal decomposition of basic zirconium sulphate precursor in the temperature range of 800°–1100°C has been used for producing ZrO_2 powder [233]. Precipitation of the precursor basic zirconium sulphate was effected by adding concentrated H_2SO_4 (~ 98%) to a $ZrOCl_2$ solution in the molar ratio of zirconium : sulphate ion = 5 : 1.6–2.8 under stirring at 90°C followed by adjusting the pH at 1.4–1.5 with NH_4OH solution. During heat-treatment from room temperature to 1100°C, the precursor released SO_3 at about 600°–650°C [233, 271]; the decomposition was complete at around 1000°C, with the formation of ZrO_2. Depending on the Zr^{4+}/SO_4^{2-} molar ratio, the particle size varied from 0.1 to 0.3 μm at 1100°C [233]. Under optimized conditions, monodisperse spheroidal particles of ZrO_2 (20 to 60 nm) were obtained (Fig. 4.6).

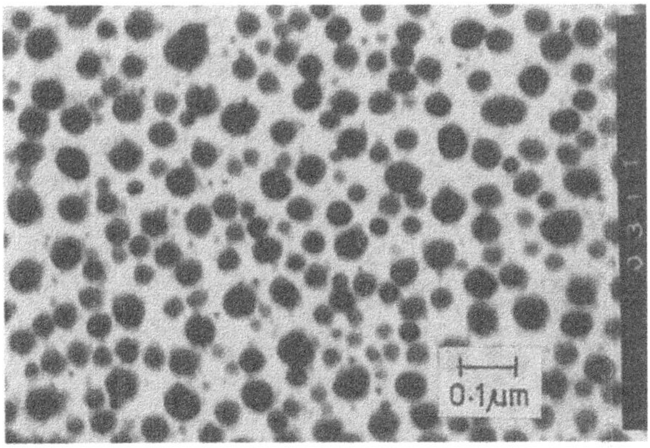

Figure 4.6 Transmission electron micrograph of monodisperse spheroidal nanoparticles of m-ZrO_2 obtained from a basic sulphate precursor [233].

Basic zirconium sulphate with spherical morphology was prepared by homogeneous precipitation from a solution containing zirconium sulphate and urea [272]. The nature of the particles was found to depend on the SO_4^{2-}/Zr^{4+} and urea/Zr^{4+} molar ratios and the cooling rate of the solution. Only the SO_4^{2-} could produce the spherical particles while NO_3^- and Cl^- ions gave rise to irregular particles. The precipitation of the basic sulphate occurred at 85°C. At 800°C, m-ZrO_2 was obtained after decomposition of the sulphate. The crystallite size of m-ZrO_2 obtained at 800°C was 70 nm.

ZrO_2 powder of ultra-high purity and submicron size was prepared by thermal decomposition of zirconium tertiary butoxide by Mazdiyasni and his associates [273]. The experiment was carried out by introducing the alkoxide into the primary vaporizor (at 190°–200°C, i.e close to the boiling point of the material) of a decomposition chamber. The vapour and the remaining hot liquid were carried by an inert gas into the secondary vaporizor where complete vaporization of the liquid occurred. At the centre of the decomposition chamber, the alkoxide vapour decomposed to ZrO_2, olefin and alcohols when it was in contact with an inert gas preheated to 325°–500°C. The powder was obtained as metastable c-ZrO_2 with an average particle size < 10 nm and transformed to m-ZrO_2 above 400°C.

Another method of obtaining stabilized fine ZrO_2 powder is the low-temperature combustion route [274]. In a typical process, MgO-stabilized ZrO_2 powder was obtained by rapid heating of a solution containing calculated quantities of $ZrO(NO_3)_2$, carbohydrazide and $Mg(NO_3)_2$ (7, 9 and 14 mol% MgO) at 350°C; the solution started boiling, and finally burnt out with a flame. The resulting powder was obtained as a voluminous foam-like, fully crystalline material. The powders with 7 and 14 mol% MgO contained t- and c-ZrO_2 respectively as single phase.

A relatively little used technique for obtaining ZrO_2 powders is the molten salt method. For the preparation of pure ZrO_2 powders, $ZrOCl_2.8H_2O$ was dehydrated under vacuum at 150°C/1 h along with the $NaNO_3$–KNO_3 eutectic mixture (i.e. molten salt mixture) [275]. At 260°C/1.5 h, the product was a mixture of m-ZrO_2 (92 vol%) and amorphous mateiral. At 400°C, both m- and t-ZrO_2 were identified. A single-phase m-ZrO_2 was obtained at 1000°C.

Using a similar but more elaborate technique, Y-doped ZrO_2 powders have also been prepared [276].

4.1.3 Titanium Oxide

Titanium dioxide, TiO_2 is generally known to occur in three polymorphic forms : anatase (tetragonal), rutile (tetragonal) and brookite (ortho-rhombic). The first two are commonly observed in natural ores and synthetic products, and will be repeatedly mentioned in the following account.

Titanium dioxide particles, in the two common forms, i.e. the low-temperature anatase and the high-temperature rutile polymorphs, are manufactured principally from natural ilmenite, $FeTiO_3$ through the following major steps [277, 278] :

(i) Digestion of ilmenite ore in sulphuric acid.
(ii) Separation of the less soluble iron sulphate, $FeSO_4.7H_2O$.
(iii) Thermal hydrolysis (under boiling) of the titanium-containing species, i.e. $TiO(SO_4)$ to obtain hydrous titania.
(iv) Washing and calcination of the hydrous oxide to obtain very fine particles necessary for use as pigments, catalysts etc.

Another important industrial method [277, 278] involves the synthesis of $TiCl_4$, which is converted to titania in the following ways :

(i) Chlorination of natural rutile to obtain $TiCl_4$.
(ii) Purification of $TiCl_4$ by distillation.
(iii) Reaction between vaporized $TiCl_4$ and O_2 to obtain TiO_2.

Apart from new variants of the industrial routes, various other processes have been developed for titania powder preparation. Examples will be cited below of both of them.

The titanium oxysulphate mentioned above has been demonstrated to be a suitable precursor chemical for obtaining titania particles of controlled shape [71]. In this process, an aqueous solution of $TiO(SO_4).nH_2O$ and urea were reacted at an elevated (70°–90°C) temperature for 0.5–1 h to obtain amorphous hydrated titania with sulphate groups, i.e. $TiO_2.xH_2O.ySO_3$. The particles were either spherical (1–4 μm) or agglomerates of finer units. Calcination of the product generated the oxide particles :

$$TiO_2.xH_2O.ySO_3 \xrightarrow{400^\circ C} TiO_2.ySO_3 \xrightarrow{500^\circ-700^\circ C} TiO_2 \text{ (anatase)}$$

$$\xrightarrow{900^\circ-1000^\circ C} TiO_2 \text{ (rutile)} \qquad \text{... (4.12)}$$

The spherical shape was retained after calcination.

The other product from natural sources of Ti, i.e. $TiCl_4$, is much more thoroughly studied and widely used in laboratory investigations. Some published accounts are described below in brief. The "oxygenolysis" or oxidation reaction of pure $TiCl_4$ was studied by Suyama and Kato [279] in a tube reactor in which the chloride vapour reacted with O_2 at 850°–1150°C; the reaction was nearly complete above 750°C, with phase-pure anatase as the product. The particles had 0.25 ~ 0.7 μm size and a square platelet morphology; the size increased with the reaction temperature. Akhtar et al. [280] reported that an increase in the reactants' concentration and residence time in the reactor, as also the temperature of reaction caused an increase in the size of the TiO_2 particles. A gas mixture of $TiCl_4$–H_2–O_2 has been used recently to obtain 100–200 nm crystalline particles by laser ignition [281].

The effect of small amounts of additives/dopants on the size and morphology of TiO_2 obtained from oxidation of $TiCl_4$ has been studied by several authors. Akhtar et al. [72] used $SiCl_4$, $POCl_3$ and BCl_3 as precursors for dopants in the vapour phase reaction. Pure titania was obtained as dense, polyhedral particles (~ 0.1 μm), while on the addition of $SiCl_4$, the morphology changed to spherical. The other dopants also yielded generally spherical/rounded submicron particles. The dopants hindered the formation of rutile; this was specially true for Si. In a similar work on the effect of dopants (Si, Al, Fe, Zr), Suyama and Kato [51] observed the following :

(i) The size of the TiO_2 particles obtained with additives became $\frac{1}{2}$ to $\frac{1}{5}$ of that obtained from a pure system, e.g. from ~ 0.1 to 0.05 μm in case of $FeCl_3$.

(ii) The particle shape could be changed from square to rounded to spherical depending on the additive. The least change was observed with $ZrCl_4$.

Vemury and Pratsinis [282], who used a laminar diffusion flame reactor with Ar as the carrier gas, confirmed the role of Si in decreasing the particle size and hindering the formation of rutile; a nearly reverse role was recorded for Sn and Al.

Aerosols obtained from citrate solutions of $TiCl_4$ generated ~1 μm hollow balloons [283]; interestingly, anatase and rutile could be obtained in different proportions (including phase-pure states) by varying the experimental parameters.

Bauer and Tomandl [284] have used $TiCl_4$ in an emulsion using a non-ionic surfactant and paraffin oil; the particle size varied in the range 0.1–2 μm, but a narrow size distribution was shown to be possible.

The tetrachloride of titanium has also been used for the preparation of nanoparticles of anatase (15–30 nm) and rutile (40–60 nm) by the calcination of products of a microemulsion of aqueous $TiCl_4$. The precipitation was caused by a second microemulsion of ammonia solution [285].

Hydrothermal synthesis of TiO_2 powder has been accomplished [286] with Ti metal as the starting material. The reaction conditions were 200°–700°C, 10–150 MPa and 0–120 h; the envisaged reaction steps were :

$$Ti + 2H_2O \rightarrow TiO_2 + 2H_2 \qquad \ldots (4.13)$$

$$Ti + \tfrac{x}{2} H_2 \rightarrow TiH_x \; (x \sim 1.9) \qquad \ldots (4.14)$$

$$TiH_x + 2H_2O \rightarrow TiO_2 + (2 + \tfrac{x}{2}) H_2 \qquad \ldots (4.15)$$

With 100 MPa pressure operative for 3 h, almost phase-pure rutile was obtained.

In a recent work [287], Ti particles were oxidized with the assistance of H_2O_2 to obtain a gel at room temperature. The gel was hydrothermally treated at 140°C. Interestingly, anatase was obtained as the reaction product in distilled water, while HNO_3 solution yielded rutile crystals. The particles were extremely fine (max. 30 nm) and acicular in morphology.

Alkoxides of titanium have been extensively used for the preparation of mostly spherical, uniformly sized, amorphous powder which was calcined to obtain anatase, rutile or their mixtures [14, 50, 60]. The reactions involved controlled hydrolysis of the alkoxide (e.g. ethoxide, isopropoxide) dissolved in an alcohol :

$$Ti(OC_2H_5)_4 + 4H_2O \rightarrow Ti(OH)_4 + 4C_2H_5OH \qquad \ldots (4.16)$$

$$Ti(OH)_4 \rightarrow TiO_2.xH_2O + (2 - x) H_2O \qquad \ldots (4.17)$$

Spherical particles (~ 0.1–1 μm) were obtained when the water/alkoxide molar ratio exceeded 2.5. The process also allowed for doping with suitable cations [59]. The mean size of the spheres could apparently be kept low by using alkoxides of higher alcohols. Such spherical, amorphous powders were generally agglomerated unless prepared from very dilute alkoxide solutions. To avoid this, addition of hydroxypropylcellulose (HPC), a surfactant [60] or ultrasonic dispersion in ammonia water [14] has been proposed. Monodisperse titania particles have also been prepared by alkoxide hydrolysis using HPC in acetonitrile-octanol-alkoxide solutions [288].

Several different (though somewhat related) techniques have been developed to convert the above batch modes of powder preparation to continuous modes. Such processes are characterized by (a) continuous flow reactors, glass bead packed bed and a static mixer [289]; (b) a stirred tank reactor system with a static mixer [290] and (c) a reactor equipped with a 100 m long tubing through which a pre-mixed solution is passed to allow for the ageing period required by the spherical, unagglomerated particles to grow in size [291].

Amorphous nanoparticles (6–60 nm) of titania have been recently synthesized [292] by causing partial hydrolysis (water/alkoxide molar ratio 1 : 1) of Ti-tetraisopropoxide dissolved in a large excess of isopropanol. Anatase crystallized at ~300°–350°C, which was almost totally converted to rutile by 615°C.

The hydrolysis reaction has also been carried out in the vapour stage by passing Ti-tetraisopropoxide and water vapour via helium into a tubular flow reactor [293] and allowing them to react at a maximum of 510°C when only anatase was obtained; the process could be operated in a continuous mode. Vapour-phase hydrolysis to obtain anatase has also been reported by Kominami et al. [294].

Oxidation of titanium metal was used for the preparation of nanoparticles of TiO_2 and their in-situ compaction [183]. The metal was evaporated in an ultrahigh vacuum gas condensation chamber under H_2 atmosphere, and allowed to deposit on a cold-finger. Passing of oxygen caused the oxidation of the metal particles to TiO_2. Various other oxidation techniques have been reported [295].

4.1.4 Magnesium Oxide

Magnesium oxide, MgO (periclase) crystallizes in the cubic system. The major routes for obtaining MgO in a commercial scale are (a) thermal decomposition of natural magnesite, $MgCO_3$, and (b) precipitation of $Mg(OH)_2$ from dissolved salts of magnesium in sea water (or fresh water brine), and calcination of the precipitate [296].

The recovery of MgO from sea water magnesia started around the early seventies. One of the earliest accounts [297] describes the process which includes the following steps :

(i) The sea water is treated with H_2SO_4 to decompose soluble bicarbonate :

$$Ca(HCO_3)_2 + H_2SO_4 \rightarrow CaSO_4 + CO_2 + H_2O \qquad \text{... (4.18)}$$

(ii) The treated water is aerated to remove CO_2.

(iii) The CO_2-free water is passed into large reaction tanks where it is mixed with hydrated lime or $MgO.Ca(OH)_2$ slurry along with $Mg(OH)_2$ seeding under stirring :

$$MgO.Ca(OH)_2 + Mg^{2+} \rightarrow MgO.Mg(OH)_2 + Ca^{2+} \quad \text{... (4.19)}$$

$$Ca(OH)_2 + Mg^{2+} \rightarrow Mg(OH)_2 + Ca^{2+} \quad \text{... (4.20)}$$

The precipitated hydroxide is flocculated with a suitable agent, and washed; all soluble impurities are apparently removed. The filter cake thus obtained is calcined at 750°–950°C.

Various solution phase techniques have been employed for the synthesis of MgO. One of the early accounts is from Brown [298], describing a method of synthesis of a high-purity (99.999%) MgO powder. The steps involved were :

(i) Preparation of a Mg-nitrate solution by dissolving metallic Mg (99.97% pure) in nitric acid under boiling.

(ii) Precipitation of the hydroxide by the addition of ammonia solution under boiling.

(iii) Addition of a purified ammonium oxalate solution, also under boiling, to obtain magnesium oxalate.

(iv) Conversion of the oxalate to MgO at 500°–1300°C. The process was also capable of preparing doped powders, and apparently had a very general applicability.

Guilliatt and Brett [299] studied the powder formation of MgO from the metal and the effect of certain impurities on the process. Magnesium nitrate hydrate crystals were synthesized by dissolving Mg metal in a minimum amount of HNO_3 under boiling. A hydroxide was obtained by the addition of ammonia into an aqueous solution of the nitrate. The washed, filtered and dried hydroxide was calcined under vacuum up to 1100°C for the decomposition to MgO. The important observation was that less than 1 atom% Mn or Fe could cause a drastic fall in the specific surface area of the powder above around 700°C, e.g. from 183 m^2/g for the pure powder to 98 m^2/g for a powder with 0.80 atom% Mn at 900°C.

Evaporative decomposition of 1M solution of the acetate, chloride, nitrate or sulphate of Mg has been studied by Gardner and Messing [300] for the synthesis of MgO. Droplets of 10 μm size were atomized with a two-fluid atomizer into a furnace kept at 500°–1000°C. The acetate salt was found to decompose with relative ease and crystallized well at 750°C into unagglomerated submicron powders. The inorganic salts underwent incomplete decomposition under similar experimental conditions, and led to ~5 μm agglomerates. In case of nitrate-derived powders, a full decomposition of the salt was achieved only at around 1000°C.

Magnesium oxide powders have been prepared [301] by the hydrothermal conversion (350°C, 16 MPa) of 99.99% Mg metal turnings to obtain submicron platelets. Doping of aluminium was carried out by the addition of Al-propoxide dissolved in trichloroethylene in a $Mg(OH)_2$–CH_3OH suspension under stirring. Single-phase MgO was obtained after calcination at 1000°C for 2 h.

The synthesis of MgO by a sol-gel method has been described recently by Portillo et al. [302]. In this process, HCl, CH_3COOH or NH_4OH was added in small quantities in a solution obtained by the refluxing of Mg-ethoxide with water and excess ethanol. Refluxing was continued till gel formation, which required the least time (~ 0.5 h) with NH_4OH. The gel was dried at 70°C/12 h and calcined in air at 300°–900°C for 4 h for obtaining the powders. All the powders contained Mg-deficient particles with the average size in the range 63–150 nm at 900°C; the specific surface area was 10–15 m^2/g. The powder was considered suitable for catalytic applications.

Various vapour phase reactions have been used for the synthesis of MgO. Thus, Nishida et al. [62, 303] described a process in which magnesium metal of 99.5% purity was vaporized by heating above its boiling point in an electric furnace. The vapour thus produced was injected through a nozzle into an oxidation chamber, where it came in contact with a stream of dehumidified air entering the chamber through a second nozzle; the reaction formed a diffusion flame. The obtained MgO particles (0.01–0.2 μm) of high (99.98%) purity were collected in a bag filter. Depending on the reaction temperature (which could go above the melting point of MgO, i.e. ~ 2800°C), two shapes were dominant : spherical and cubic.

In a similar work, Watari et al. [304] generated 0.005–0.4 μm single crystals of MgO (cube or rectangular parallelopiped) through a vapour phase reaction. Magnesium vapour was allowed to react with oxygen passed into the reactor by a mixture of Ar and N_2 as the carrier gas. The particle size of MgO was found to decrease with an increasing ratio of $[O_2]/[Mg]$ and increasing reaction temperature (890°–1130°C).

4.1.5 Silicon Dioxide

Silicon dioxide (SiO_2) is the only compound selected in this Chapter for which preparative techniques will be discussed for obtaining non-crystalline (amorphous or glassy) particles as the final product. Silica has many polymorphs, the most abundantly available and extensively exploited being the low form of quartz (or α-quartz), which is also the thermodynamically stable phase at room temperature. Low quartz is one of the raw materials for non-oxide powders (e.g. SiC), and is processed for this purpose by the comminution of various natural sources. For many solution techniques, a conventional precursor is a silica sol, prepared from sodium silicate solutions via the polymerization of silicic acid, and containing particles of colloidal dimensions (a few nm to a few hundred nm). Another such conventional product is fumed silica, which is a vapour phase reaction product.

Since the recent re-evalution of the publication of Stöber et al. [69] on silica particles (discussed below), many new investigations have been reported on the preparation of porous and dense amorphous silica powders. Some of them will be discussed here, starting from the seminal paper.

Stöber et al. [69] reported the synthesis of spherical, monosized, monodisperse amorphous silica powder from silicon alkoxide solutions in 1968. The basis of the experimental work was mixing a silicon alkoxide (ethoxide or methoxide, or also the reaction products of $SiCl_4$ with higher alcohols) with a solution containing one or more of methanol, ethanol, n-propanol and n-butanol, alcoholic ammonia solution, ammonium hydroxide, and also water for the hydrolysis reaction. After a period of incubation (1–5 min), spherical silica particles were precipitated out. Ammonia was found to be the promoter of sphere formation, as ammonia-free reactions yielded only irregular flocs. The spherical particles were up to ~ 2 μm in diameter, higher alcohols yielding large sized particles. The effects of different solvents on the particle morphology and size have been examined [e.g. 70] for the Stöber technique. Table 4.2 shows the nature of the interdependence between the type of solvation of the ethoxide molecules (TEOS) and the nature of particle growth under identical conditions; Fig. 4.7 shows particles synthesized with methanol as the solvent.

Table 4.2 Effect of solvents on the nature of ethoxide-derived silica particles [70]

Solvent	Incubation period (sec)	Average particle diameter (μm) and particle morphology
Methanol	144	0.22; spherical
Ethanol	65	0.43; ,,
1-Propanol	53	1.0; ,,
2-Propanol	80	0.88; ,,
1-Butanol	25	Non-spherical agglomerate
i-Butanol	0	,, ,,
Acetone	150	,, ,,
1, 4 Dioxan	184	,, ,,

It was thus observed that for spherical particles, the incubation period increased, while the particle size decreased with increasing polarity of the solvent.

Yoshizawa et al. [305] have used a seeding technique to control the particle size of spherical silica from 0.2 to 1.6 μm, thus demonstrating another aspect of versatility of the technique. A continuous production process for the powder has been discussed by Giesche [306].

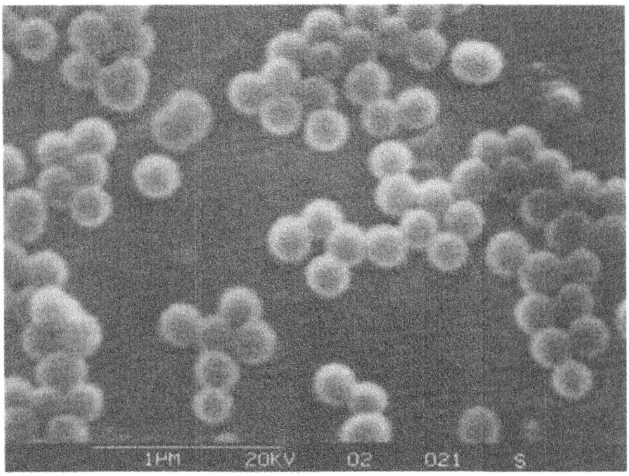

Figure 4.7 Monodisperse spherical silica particles obtained by the Stöber technique with methanol as solvent [70].

Bogush and Zukoski [307] have made an elaborate study of the Stöber technique and the possible growth mechanism of the spheres. They found that reproducible synthesis of the monosized spheres was possible over a size range of 15–700 nm by varying the conditions of synthesis. The chemical parameters and the powder size were correlated in the following way :

$$d = A\,[H_2O]^2 \exp(-B[H_2O]) \qquad \ldots (4.21)$$

where $A = [TEOS]^{-(1/2)}\,(-1.042 + 40.57\,[NH_3] - 9.313\,[NH_3]^2)$
and $B = 0.3264 - 0.2727\,[TEOS]$; d is expressed in nm.

Bogush and Zukoski further inferred that the nucleation time was not short, as conventionally supposed in such cases, but took place through a considerable part of the reaction time. A nucleation-aggregation model was suggested by these authors. The aggregation mechanism was further discussed by Giesche [306].

While the Stöber technique dealt with particle synthesis under highly basic conditions, other investigators have identified regimes in alkoxide-water-acid systems where spherical, amorphous silica particles can be synthesized. Kawaguchi and Ono [308] obtained spherical silica particles from the low-water, immiscible region of the ternary system tetraethyl orthosilicate (TEOS)-ethanol-water in presence of a fixed amount of HNO_3. The generation of spherical particles has been attributed to emulsion formation in the two-phase region of the above system. A regular emulsion system with tetramethyl orthosilicate (TMOS)-methanol-water solutions and various surfactants has been reported by Yamashita et al. [309] to generate ~ 8–16 μm microporous particles.

In another investigation on emulsification [164], an acidified water/organic solvent emulsion with 0.1 vol% of a non-ionic surfactant was used as the base, in which TEOS was added. After a period of development of turbidity, microspheres formed and settled at the bottom of the container. Such microspheres, after calcination at 450°C, exhibited high specific surface area (200–400 m^2/g). A somewhat similar work has been reported recently by Butler et al. [310].

Spherical particles were also reported by Karmakar et al. [311] from the system TEOS-acetic acid-water with specific compositional ranges (molar) TEOS : CH_3COOH : H_2O = 1 : (0.5 – 4.0) : (1.5–5.0). The spheres were up to ~100 μm in size, and were almost fully dense without calcination.

Ding and Day [312] have described a sol-gel method by which relatively large, glassy silica spheres could be prepared on a regular basis. The method consisted of passing a partially hydrolyzed, low-water, acid-catalyzed alkoxide solution through a droplet generator operated by a vibratory system. The droplets thus generated were converted to gel microspheres by heating them to < 250°C. After calcining the product at 800°C, densified glassy microspheres of ~ 60 μm size were obtained. The spheres were nearly monosized.

The most important vapour-phase reaction involving the preparation of high-purity silica powders is the oxidation of $SiCl_4$ vapour (also called "flame hydrolysis") in an oxy-hydrogen flame. As mentioned in Section 3.5.2, this is a commercial process for the production of "fumed" silica. The reactions involved are [313] :

$$2H_2 + O_2 \rightarrow 2H_2O \quad \text{... (4.22)}$$

$$SiCl_4 + 2H_2O \rightarrow SiO_2 + 4HCl \quad \text{... (4.23)}$$

$$SiCl_4 + 2H_2 + O_2 \xrightarrow{1800^\circ C} SiO_2 + 4HCl \quad \text{... (4.24)}$$

Molten spheres of silica are thus produced, which collide with one another to form, on cooling, glassy aggregates 0.03 μm or less in size. The surface area varies in the range 100–400 m^2/g. A powder of very high chemical purity (all metals below 5 ppm or even less) is obtained by this process. A relatively impure fumed silica, finding wide applicability in the refractories industry, is also obtained by vapour-phase reactions in the ferrosilicon/silicon metal producing industries [e.g. 314]. The reactions include the reduction of SiO_2 to SiO vapour, which comes in contact with oxygen at the top (mouth) of the furnace and condenses as submicron, spherical particles of SiO_2.

4.1.6 Rare Earth Oxides

Rare earth oxides are in general high temperature materials and some of them find wide applications in ceramics, e.g. as dopants or additives in zirconia or silicon nitride ceramics, as will be seen elsewhere in this book. Rare earth oxide ceramics are currently finding new applications [315]. In this Section, two examples (yttrium and cerium oxides) have been selected out of the large group to acquaint the readers with the different techniques that have been generally used in the synthesis of rare earth oxide powders.

Yttrium Oxide

Precursors of Y_2O_3 have been obtained by the simple precipitation of the hydroxide from yttrium nitrate or chloride solution by the action of NH_4OH solution [316, 317]. The precipitates were voluminous and flocculated; washing with water and ethanol, and calcination at 700°C, converted them to irregular chunks of the oxide with low surface area. In a forced hydrolysis reaction at 90°C in presence of urea, the solid product consisted of spherical particles of the basic carbonate, transforming to Y_2O_3 spheres (< 0.5 μm) of narrow size distribution and a specific surface area of ~ 5 m^2/g after calcination at 700°C [317].

Akinc and Celikkaya [318] prepared an water-in-oil emulsion by the dropwise addition of a yttrium nitrate solution into a stirred bath of toluene containing small quantities of non-ionic surfactants. For submicron particles, ultrasonication was used. Triethanolamine was added to the emulsion to obtain precipitation of yttrium. The precipitate was centrifuged, washed in water, dried at 100°C and calcined at 700°C/5 h.

A spray-dried yttrium oxide powder was reported by Hours et al. [319]. A precipitate was first prepared through the interaction of $YCl_3.6H_2O$ with NH_4OH solution. The precipitate was peptized by the action of an acid (generally HCl). The sol thus obtained was spray-dried. The powders calcined above 600°C crystallized to only Y_2O_3. The special character of the spray-dried agglomerates (~5–15 μm) was their extreme porousness, indicating easy crushability.

In a recent work, hydrated Y-nitrate solutions in n-propanol were sprayed by an electrospray atomizer suitable for obtaining sufficiently small droplets for submicron particles. Spheroids of up to ~200 nm were obtained at 550°C [320].

Various special techniques have also been used for the synthesis of Y_2O_3. Micheli [321] utilized the double decomposition reaction between a water-soluble ammonium polyacrylic salt and a Y-salt solution (the technique is valid for many other multivalent cations) to produce a metal polyacrylate precipitate and a soluble ammonium salt. The yttrium polyacrylate thus obtained was calcined at above 1000°C to convert it to soft agglomerates of Y_2O_3.

Minagawa and Yajima [322] prepared a precursor salt, $NH_4Y(C_2O_4)_2.H_2O$ by reacting NH_4OH solution with Y-oxalate solution or a Y(III) slurry with oxalic acid. The oxide was obtained by thermal decomposition of the salt at 750°C.

Both solid and solution combustion techniques have been used by Ekambaram and Patil [323] for Y_2O_3 powders. In the solid combustion technique, yttrium hydrazine carboxylate complex, $Y(N_2H_3COO)_3.3H_2O$ was pyrolyzed in air at 600°C/ 1 h. The complex was also rapidly heated with NH_4NO_3 or NH_4ClO_4 at 300°C. Direct pyrolysis resulted in the finest powders (0.04 μm) and agglomerates (1.10 μm), as also the largest surface area (36 m^2/g). In the solution combustion process, a saturated aqueous yttrium nitrate solution was rapidly heated with various hydrazine-based fuels to obtain fine particles (0.068 μm).

Cerium Oxide

Cerium (IV) oxide has been obtained by the forced hydrolysis (at ~ 90°C) of the cation in aqueous solutions of its nitrate and sulphate salts [324]. The solid product has been shown to be very much dependent on the starting salt, kind of impurities (if present) in the solution, solution concentration, temperature and pH. The morphology and size varied as widely as spheres (~ 0.2 μm) and discrete rods (~ 1–3 μm) by changing the experimental conditions. In addition, the product could be the desired oxide or an oxy-salt. Matijevic and Hsu [325] showed that Ce (III) ions precipitated as crystalline $CeO(CO_3)_2.H_2O$ from a urea solution. Cerium oxide powders, on the other hand, were prepared by the homogeneous precipitation from a Ce(III) nitrate solution in presence of hexamethylenetetramine [326].

Both hydrothermal [327] and "solvothermal" [328] techniques have been used recently to obtain pure or doped (Mg, Ca, Sc, Y, Nd) microcrystals of CeO_2. The starting material for hydrothermal synthesis was a precipitate from a nitrate solution or a colloidal suspension of $Ce(OH)_4$ in water. The precipitation took place at pH ~ 9–10, a temperature up to 350°C and pressure, 10–15 MPa. In case of solvothermal synthesis, the $Ce(OH)_4$ suspension was made in ethanol. Nanoparticles of high specific surface area (~ 100m^2/g) have been reported. Similar work on hydrothermal synthesis has been reported by Tani et al. [329].

The most recent work has been reported by Hirano and Kato [315] who prepared CeO_2 powders from cerium (III) nitrate, cerium (IV) sulphate and cerium (IV) ammonium sulphate, starting with precipitation from the salts through the action of ammonia. The precipitates with excess ammonia were treated hydrothermally at 120°–200°C for 5–40 h.

4.1.7 Zinc Oxide

Commercially, zinc oxide (ZnO) is known to be produced in two different ways [330, 331] :

(i) The so-called French process, in which metallic zinc is first vaporized, and the vapour is led to a combustion chamber where it is oxidized in air to form ZnO particles.

(ii) The so-called American process, in which zinc sulphide ore concentrates are oxidized by roasting, mixed with anthracite coal and smelted in a flatbed furnace. Vapours of metallic zinc are produced, which are burned under controlled conditions and led to the bag house where the oxide particles are collected.

Apart from the above, various laboratory methods have been developed for the synthesis of ZnO powders, some of which are described below.

Fabrication of varistors being one of the major applications of ZnO, many investigations in powder preparation have been directed towards this goal. In one such account, Lauf and Bond [87] reported the preparation of individual hydroxide precipitates of Bi^{3+}, Sb^{3+}, Co^{2+}, Cr^{3+} and Mn^{2+}, along with Zn^{2+}, which were mixed together and dispersed in water by peptization. The chloride ions (in case of Mn and Sb) were carefully removed. The colloidal sol obtained in this way was gelled by the evaporation of the water and the gel calcined at 250°C to obtain submicron, doped ZnO particles.

Varistor compositions were also prepared by a urea process [332]. The dopants in the form of Bi_2O_3, CoO and MnO were dissolved in HNO_3; the oxide of another dopant, Sb^{3+} was converted to the citrate in molten citric acid, and the product diluted with water and mixed with the nitrate solution. Granular urea was added to the solution, which was heated at 250°C to cause the formation of amides. The precursor thus obtained was calcined at 675°C so as to decompose it and obtain the powder.

In another investigation [333], the starting compounds ZnO, Bi_2O_3, $SbCl_3$, $CoCl_3.6H_2O$, $CrCl_3.6H_2O$ and $MnCl_2.4H_2O$ (for obtaining 0.5–1 mol% of the doping oxides in ZnO) were dissolved in HCl, with excess acid to prevent the hydrolysis of Bi and Sb. The mixed solution was added to a large excess of a diethylamine solution for co-precipitation as hydroxides at a final pH of ~ 8–9. The precipitate was filtered, washed with a relatively dilute solution of diethylamine and dried at 110°C. Calcination at 950°C /1 h yielded ZnO as the major phase, and in addition, a spinel phase and an unidentified Bi-rich phase.

A two-step reaction scheme was followed by Haile et al. [334]. The first step was the synthesis of a zinc hydroxy salt precursor, prepared by one of the two ways given below :

(i) $ZnSO_4.7H_2O + NH_4OH \rightarrow [xZnSO_4.yZn(OH)_2.zH_2O]$ ... (4.25)

(ii) $ZnCl_2 + NH_4OH \rightarrow [xZnCl_2.yZn(OH)_2.zH_2O]$. ... (4.26)

In both cases, the salt was first dissolved in pure water to obtain a 1M solution, with NH_4OH slowly added to it under stirring; the precipitate (basic zinc salt) was washed and dried at 100°C. The second step involved (i) the reaction of these precursors with a limited amount of NH_4OH to just dissolve them, and (ii) evaporation of the solution at various temperatures above 55°C to effect precipitation of hydroxide and its conversion to ZnO. Various organic additives (e.g. methanol, ethanol, triethanolamine) were used for achieving good dispersion and particle sphericity. Addition of the necessary dopants (for varistor applications) was accomplished by precipitation from the respective salts onto the ZnO particles dispersed in methanol. Micron-sized particles of spherical and other morphologies were obtained as the end product.

The effect of organic compounds in wet chemical processes on the particle morphology of ZnO has been examined by Costa and Baptista [85]. In the synthetic process, aqueous acidic solutions of $ZnCl_2$ were fine-sprayed into an NaOH solution, or one containing (i) (mono)ethanolamine or (ii) triethanolamine for precipitation. The pH of the precipitating solution was constantly maintained at 12 by the necessary addition of NaOH. The particles thus obtained were washed with water for removing sodium and finally with acetone-toluene-acetone to remove the surface water. Both pure water and ethanol generated irregular, 19–25 nm platelets (the latter produced smaller particles); the specific surface area ranged from 18 to ~ 45 m^2/g. Ethanolamine produced elongated hexagonal prisms, while triethanolamine led to the formation of cauliflower-shaped particles; the particles were nanosized but relatively large, with low specific surface area (1 to ~ 19 m^2/g).

Recently, Zhong and Matijevic [335] have reported the synthesis of crystalline colloidal particles of ZnO by a forced hydrolysis reaction involving zinc nitrate in aqueous solution and a triethanolamine solution (double-jet). The two were introduced simultaneously in a constant flow into a reactor containing water at 90°C under stirring. The particles varied in size from submicron to more than 1 μm with decreasing flow rate of the reactants.

It has been already mentioned that depending on the conditions, ZnO powders can be synthesized in various morphologies. Detailed studies on the control of conditions of synthesis are available, some of which are described below.

Fujita et al. [61, 83] used low-concentration aqueous solutions of Zn-nitrate and chloride solutions with hexamethylenetetramine (HMT) for homogeneous precipitation of ZnO (solution pH = 1 –3). HMT hydrolyzed first, generating ammonia :

$$(CH_2)_6N_4 + 6H_2O \rightarrow 4NH_3 + 6HCHO \quad \text{... (4.27)}$$

which was followed by the precipitation of ZnO. The optimum concentration ranges of the chemical inputs at different pH conditions were documented. It was also shown that the particle morphology drastically changed from globular to acicular when the concentration of HMT increased. The crystal shapes (globules, needles, rods) were found to have a complex dependence on the different chemical parameters. With the optimum proportions of the Zn-salt and HMT, NH_4Cl and NH_4NO_3 suppressed the formation of globules, while formaldehyde (HCHO) favoured this morphology.

Another investigation of a similar nature by Sakka et al. [84] showed in addition to the above that in an aqueous solution of Zn-nitrate (0.05 M)–HMT (0.1 M)–HNO_3 at 90°C, the initial precipitate was globular, but gradually changed to a rod-like morphology after 0.3 h of ageing in the hot solution. The yield was about 60% after 4 h, which increased to nearly 80% by increasing the HMT concentration to 0.2M. The particle size varied from submicron to several microns, and the maximum (and the most reliable) value of the specific surface area was ~ $9m^2/g$.

Hydrothermal conditions (open and closed) have been employed to achieve oxidation of fine Zn metal particles [286]. Oxidation at 300°C under 100 MPa pressure was reported to have been complete after about 1 h, without any intermediate product. The crystallites of ZnO were 60–160 nm at 500°C. Hydrothermal decomposition of Na_2Zn–EDTA at ~350°C/> 10 MPa has been studied [336]; smooth, needle-like crystals of ZnO were obtained within short reaction times (up to ~1 h).

Another simple method [337] consisted of refluxing a diethylene glycol solution of Zn-acetate dihydrate at 180°C when well-crystallized ZnO particles were obtained. The particles were spherical, monodisperse, porous, 0.2–0.4 μm in size with a narrow size distribution and a specific surface area as high as 60 m^2/g. The calcined (500°, 730°C) particles were non-agglomerated, 20 nm in crystallite size, having a specific surface area of 5 m^2/g.

Alkoxide precursors have also been used for the synthesis of ZnO. Vogel et al. [338] obtained ZnO powder simply by the hydrolysis of a dilute (0.2 M) solution of zinc ethoxide in dry ethanol by its contact with a water-ethanol mixture. Precipitation was immediate, and the wet product thus obtained was subjected to several cycles of re-dispersion and settling in ethanol. The particles had a mean size of 0.04 μm. Hydrolysis of ethylzinc-t-butoxide has been used as a route for obtaining generally submicron, equiaxed ZnO particles [339].

Bis-(acetylacetonato)-zinc (II) as a precursor of ZnO was prepared by Kamata et al. [340] by a simple method. A methanolic solution of this compound was added slowly to deionized water under vigorous stirring at 25° and 80°C. The immediately precipitated product was further aged for 1 h under agitation, filtered, dried in methanol and vacuum dried at 50°C. Both the products were crystalline, single-phase ZnO (better crystallized at 80°C). While the product of 25°C was irregular, that of 80°C was rod-shaped (~ 500 nm length, ~ 100 nm cross-section).

Ultrafine ZnO of different morphologies was reported from vapour phase oxidation of Zn [341]. While globular particles retained their size to ~ 0.05–0.07 μm with increase in the reaction temperature from 850°C to 1100°C, needle-like particles grew rapidly in size to more than 0.2 μm under similar conditions; the latter grew further with an increase in the concentration of zinc in the reaction vapour. Zinc oxide crystals were also obtained by oxidizing Zn vapour at 850°–1070°C [342]. The submicron crystals were spherical in morphology at the lowest size and granular when increased in dimensions; the largest crystals were tetrapod-like twins.

4.1.8 Mullite

Mullite is known to be the only stable "compound" in the phase diagram (at atmospheric pressure); broadly, its chemical formula ranges from $3Al_2O_3.2SiO_2$ to $2Al_2O_3.SiO_2$, indicating a solid solution [343], though highly alumina-rich mullite has also been reported [344]. The following discussion on mullite synthesis essentially centres round the 3 : 2 composition indicated above.

The composition range of mullite indicates predominance of Al_2O_3 over SiO_2, so in the early investigations, clay minerals were mixed with alumina or alumina-rich minerals and fired at high temperatures (~ 1700°C). Other solid-solid reactions involved gibbsite, diaspore, α-Al_2O_3 and the silica polymorphs (mainly quartz) and silica glass. However, in all such reactions, both time and temperature were unacceptably high, and the impurities of the natural raw mateirals led to undesired effects. The above, however, is not to indicate that solid-solid reactions for obtaining mullite *ceramics* is an outdated and rejected method. Investigations are continuing for obtaining the optimum conditions of synthesis [345, 346].

Extensive work has been carried out on the preparation of mullite powders using sol-gel procedures. Considering the variations in starting materials, the methods can be divided into the following major groups.

I. Methods based on particulate (colloidal) sols of Al and Si.

II. Methods based on inorganic Al-salts or colloidal sols and Si-alkoxides or colloidal silica sols, but not a combination of colloidal sols.

III. Methods based on Al-alkoxides and Si-alkoxides.

In the following, some examples from each of these methods will be discussed.

Method I generally involves dispersion of colloidal particles, i.e. fumed alumina, AlO(OH), γ-Al_2O_3 etc. of high surface area in an aqueous medium or use of a commercial alumina sol in combination with a dispersion of fumed silica or silica particles prepared from alkoxides or Na-silicates; the latter yields impure starting materials, not suitable for exacting applications.

An example of Method I is provided by the work of Ghate et al. [347] who slowly added an ammonia-stabilized silica sol into a 7 wt% suspension of fumed alumina in water at pH ~ 5. Dropwise addition of ammonium hydroxide increased the pH to 6–7 and caused gelation. The gel pieces, when dried (120°C), crushed and sieved, crystallized into mullite on heating; a complete mullitization required at least 1400°C/20 h. The particles were submicron in size, globular (as against the common needle-like morphology) and 99.9% pure. Similar results were reported by Sacks and Pask [348] who obtained a gel by pH adjustment or liquid evaporation from a mixture of colloidal γ-Al_2O_3 and commercial coloidal silica sols.

An improvement over the above results was reported by Ismail et al. [349] who, using similar starting materials, obtained complete mullitization (1 μm flakes) after only 1 hour at 1400°C. Mizuno and Saito [350] prepared a hybrid colloidal sol by dispersing boehmite and fumed silica in water so as to obtain 71.8 wt% of equivalent Al_2O_3 and 28.2 equivalent wt% SiO_2 (i.e. stoichiometric $3Al_2O_3.2SiO_2$). The gels prepared by thermal treatment of the sols were dried and heated at 1350°C/3 h; phase-pure mullite particles were obtained.

Recently, Lieu and Wilcox [351] employed an emulsion process with water extraction (by n-butanol) in which a mixture of a boehmite sol and a commercial silica sol was used; micron-sized hollow spheres were obtained.

A typical example of Method II is the elaborate work of Mizuno and Saito [350] who used 6–8 wt% (equivalent Al_2O_3) solutions of one of the sulphate, nitrate or chloride salts along with fumed silica to obtain a number of sols (of 3 : 2 composition); one half of each of them was neutralized to pH 7 by the addition of NH_4OH solution. The sol obtained from Al-sulphate and fumed silica, and not subjected to neutralization, produced phase-pure mullite by a heat-treatment of the corresponding gels at 1350°C/3 h, while the others generally yielded α-Al_2O_3 and cristobalite (SiO_2) or both as impurities. The nitrate-derived sol, without neutralization, produced single-phase mullite, but only at 1650°C. Apparently, an increase in the pH caused precipitation from the salt solution, leading to the crystallization of undesired phases.

Jaymes and Douy [352] spray-dried sols made from Al-nitrate and TEOS to obtain 2–10 μm spheres. The products from sols without ageing yielded phase-pure mullite at 975°C.

Similar work has been reported by other workers [e.g. 353, 354] when the source of SiO_2 was TEOS or ethyl silicate. Kanzaki et al. [354] spray-pyrolyzed the mixed sol to obtain amorphous, ~ 1 μm hollow spheres, which directly transformed to well-crystallized 3 : 2 mullite at around 1000°C. Moore et al. [355], on the other hand, prepared solid submicron spheres of mullite by aerosol decomposition of an Al-nitrate/fumed silica sol.

Method III, using only alkoxides, has been widely studied. Early work in this line was reported by Mazdiyasni and Brown [356] who prepared very high surface area, needle-like, highly pure mullite crystals from a mixture of isopropoxides refluxed in isopropanol after calcining the solid product at around 1200°C.

Mitachi et al. [357] prepared mixed benzene solutions of Al-isopropoxide and TMOS, apparently to control hydrolysis; a substantial quantity of mullite was obtained at 1000°C (with a little "spinel" phase). At 1250°C, only mullite with 71.6% Al_2O_3 was obtained. Suzuki et al. [358] refluxed an acid-catalyzed TEOS solution in ethanol and water, and an Al-isopropoxide solution in isobutanol separately, and mixed them for reaction for 24 h; an amorphous powder was finally obtained. When calcined at 1200°C, this precursor yielded phase-pure mullite particles of average size 23 nm and specific surface area 83 m^2/g. Jeng and Rahaman [359] recently showed that crushing of mullite gel powders would be facilitated if the gels were hypercritically dried.

Detailed investigations on the products of Method III have been carried out by Schneider and associates who used Al-s-butoxide and TEOS [360, 361]. Figure 4.8 shows a product obtained by Schneider following Method III. Paulick et al. [362] reported their success in the synthesis of 100% 3 : 2 mullite only at 550°C/1 h by calcining a product of hydrolysis of $Al(OC_3H_7)_3$ and TEOS under conditions of acid catalysis.

The hydrothermal route has been employed extensively for obtaining mullite particles. Somiya et al. [363] treated alkoxide-derived precursors of alumina with TEOS under hydrothermal conditions (300°–600°C and 10–100 MPa pressure). Three different processing routes were followed :

(a) hydrothermal treatment of a benzene solution of Al-isopropoxide and TEOS;

(b) hydrothermal treatment of a washed pseudoboehmite phase obtained via route (a);

(c) hydrothermal crystallization of a calcined (600°C) precipitate obtained via hydrolysis of an isopropanol solution of Al-isopropoxide and TEOS.

Routes "a" and "b" were not found to be promising for pure-phase mullite synthesis. In route "c", the calcined product yielded essentially mullite at 600°C/20 MPa after

2 h. With increasing time of reaction, the quantity of mullite increased, but an aluminosilicate hydrate phase also appeared in small quantites with prolonged ageing.

Figure 4.8 Transmission electron micrograph of a mullite precursor obtained from tetraethyl orthosilicate (TEOS) and aluminium sec-butoxide (kind courtesy : Dr. H. Schneider, Deutsche Forschungsanstalt für Luft- und Raumfahrt e.V., Köln, Germany).

In another investigation [364], Al-and Si-acetates were hydrothermally co-processed. Calcination of the solid product at ~ 1250°C gave rise to a phase-pure mullite (3 : 2). Silicon acetate has also been used [365] as the source of silica in conjunction with Al-nitrate or Al-sulphate for preparing gels. In the latter case, the 3 : 2 composition yielded only mullite at ~1325°C/18 h. Nearby compositions generated mullite as the single or major phase, with corundum as a minor phase. The process provided for addition of dopant cations in the form of soluble salts. A citrate gel method, using citric acid and acrylamide, was used by Douy [366] for the preparation of mullite.

Hong et al. [367] employed the thermal decomposition and combustion techniques for obtaining mullite within very short periods. Stoichiometric amounts of $Al(NO_3)_3 \cdot 9H_2O$ and fumed silica were mixed and ground in presence of water. For thermal decomposition (process D), a low-urea (urea/Al-nitrate molar ratio = 1) and high-water composition was selected. For combustion (Process C), a high-urea and low-water composition was used. The slurries were put in a furnace, idling at 500°C, for 5 min. Process C produced 7–8 μm mullite particles with minor quantities of corundum and Al-Si spinel, while Process D produced spherical, 0.1–0.2 μm amorphous pre-mullite particles of high surface area. Calcination of the obtained product at a temperature as low as 950°C/2 h yielded mixtures of mullite and the Al-Si spinel phase, as also pure mullite under specific conditions of precursor chemistry. The mullite particles were ~ 0.1 μm platelets.

Among the other techniques, the reaction between $AlCl_3$ and $SiCl_4$ apparently holds promise [368]. The initial work is about three decades old, where the two chlorides were dissolved in diethyl ether and hydrolyzed with aqueous ammonia solution. However, pure mullite was not obtained below 1400°C. Hori and Kurita [369], on the other hand, prepared submicron, spherical pre-mullite powder (67–80 wt% Al_2O_3) by the chemical vapour deposition technique, using $AlCl_3$ and $SiCl_4$ vapours with N_2 as the carrier gas. The powder was calcined at 900°–1500°C; mullite was the major phase at ~1100°C, but single-phase mullite (with 71.2 wt% Al_2O_3) was obtained only above 1300°C.

4.1.9 Magnesium Aluminate Spinel

The cubic spinel ($MgAl_2O_4$) structure is the basis for a large number of AB_2O_4-type compounds with widely varying and involved crystal chemistry and crystallography. The synthesis of a part of these compounds will be discussed separately in Section 4.1.11. In this section, the commonly used procedures for the synthesis of the base compound, $MgAl_2O_4$ will be described.

One of the oldest techniques of spinel formation is solid-solid reaction, which has been employed by many workers more for understanding the kinetics than for the synthesis of a powder for ceramic applications. Thus, Bakker and Lindsay [370], who studied the synthesis of $MgAl_2O_4$ from commercial Bayer alumina trihydrate and commercial magnesium trihydrate with or without $AlF_3.xH_2O$ as a mineralizer, showed the importance of the latter in obtaining a substantial quantity of spinel above 900°C. The friable compact could be broken down to 2–3 μm particles. Beretka and Brown [371] examined the reaction kinetics of spinel formation by using large (45–53 and 90–105 μm) and fine (~ 1–3 μm) particles of MgO and Al_2O_3 up to 1500°C for obtaining high degrees of conversion.

The most well-known route for obtaining $MgAl_2O_4$ particles is co-precipitation from solutions. One of the oldest contributions comes from Bratton [372, 373] who recognized that the two metal ions showed facile precipitation of their hydroxides under two different conditions : Al-hydroxide completely precipitated out at a pH of 6.5–7.5, while Mg-hydroxide required strongly basic solutions. A mixed solution (Mg : Al = 1 : 2) of $MgCl_2.6H_2O$ and $AlCl_3.6H_2O$ prepared at 80°C was therefore poured into a NH_4OH solution at a constant temperature under stirring; precipitation occurred at pH of 9.5–10 (covering the values for both the cations) in a controlled way. The washed and dried gelatinous precipitate contained gibbsite and a double hydroxide, but was amorphous at 350°–400°C (apparently a transition stage). Above 400°C, spinel started crystallizing. The particle size of the powder increased from 0.03 μm (400°C/16 h) to 0.21 μm (1400°C/6 h), while the specific surface area decreased dramatically from 140 m^2/g (400°C/2 h) to only ~12 m^2/g (1200°C/2 h).

Co-precipitation of Mg and Al from aqueous solutions with Mg/Al ratio of 1 : 2 at pH = 10 and 60°C [374] and drying at 100°C of the precipitate thus obtained was thought to cause the formation of a hydroxycarbonate of the two metals, along with AlO(OH) and the trihydroxides. Up to 280°C, only transition phases were observed. Poorly formed spinel was detected at 400°C, which became well-crystallized at 1100°C. Spinel also formed in substantial quantities (along with some MgO) at 1100°C when the Mg/Al ratio in the precursor nitrate solution was 2 : 1. The question of formation and stoichiometry of spinels by the co-precipitation route has been discussed by Bhattacharya and Samaddar [375].

Galimov et al. [376] considered that the reaction scheme

$$\begin{aligned}&2NaAlO_2 + a NaOH + Mg^{2+} + a H^+ + n H_2O \rightarrow \\ &Mg(OH)_2.2Al(OH)_3 + (2 + a)\, Na^+ \qquad \ldots (4.28)\end{aligned}$$

could be modified to avoid the addition of acid (H^+) meant for neutralizing the excess alkali. This modified process replaced acid with an Al-salt by a complex balance of the chemical entities in the system $NaAlO_2$–NaOH–$MgCl_2$–$AlCl_3$–H_2O. Solutions of $NaAlO_2$, $MgCl_2$ and $AlCl_3$ were prepared separately, and mixed together at 70°–80°C. A white precipitate, containing some transient crystalline phases, was obtained under stirring. The precipitate was filtered, washed and dried at 25°C/150°C, and finally calcined at 1200°C/2 h. Crystallization was found to start at 280°–300°C. The spinel (97–99%) crystallized at 1200°C had a MgO/Al_2O_3 molar ratio which could vary in the range 0.98–1.03. The stoichiometry of the precursor could be controlled to adjust the Mg/Al ratio. It seems that contamination of the spinel crystals by sodium may be a problem for post-processing.

A freeze-dried mixed sulphate precursor was prepared by Nakagawa et al. [377] from an aqueous solution of Mg- and Al-sulphates, which was calcined at 900°–

1400°C/1 h. The mixed sulphate, which was already depleted in structural water content after freeze drying (from $22H_2O$ to 8–9 H_2O per molecular unit), decomposed completely and transformed to spinel above 1000°C. The apparent crystallite size was ~ 60 nm and the specific surface area was 20–30 m^2/g.

Spinel has been synthesized also by the preparation of a colloidal or polymeric sol and gelling it to obtain partly or fully amorphous gel particles; these could be calcined or compacted and sintered to obtain crystalline particles/grains. An early work in this area was carried out by Mitchell [378]. Stoichiometric quantities of (i) aluminium hydroxychloride, $Al_2(OH)_5Cl$. 2–3 H_2O in an aqueous solution and (ii) a suspension of $Mg(OH)_2$ in water were mixed; they reacted rapidly to form a gel. The gel was dried and ball-milled to a fine powder. At 900°C, the gel partially converted to spinel. Compacted material, hot pressed at 1000°C followed by heating at 1300°C, yielded well-crystallized spinel.

Sugiura and Kamigaito [379] used a mixed alkoxide, $MgAl_2(i\text{-}OC_3H_7)_8$ for the synthesis of spinel. The alkoxide was dissolved in a 1 : 1 mixture of benzene and ethanol under Ar atmosphere. Water was added dropwise under stirring, when a white powder precipitated out. Drying (90°C, vacuum) and calcination (700° ~ 1200°C) led to the crystallization of only spinel at 700°C, which became well-crystallized at 1000°C. The particle size of the spinel was 5–10 nm at 800°C, which increased to 0.1 μm at 1200°C. The specific surface area, on the other hand, decreased from 334 m^2/g at 200°C to 0.03 m^2/g at 1400°C. The alkoxide was also separately dissolved in n-hexane for hydrolysis reaction; the precipitate obtained therefrom yielded only spinel at 800°C.

A detailed experimental study of gelation/powder formation in four different systems containing Mg and Al, in seeded (1.69 wt%) or unseeded condition, has been reported by Pasquier et al. [380].

The systems were :

(i) Commercial boehmite, dispersed in water to obtain a sol and $Mg(NO_3)_2.6H_2O$ solution.
(ii) The same type of boehmite sol and ultrafine powder of MgO.
(iii) $Al(NO_3)_3.9H_2O$ and $Mg(NO_3)_2.6H_2O$ mixed solution.
(iv) Al-isopropoxide and Mg-ethoxide refluxed at 80°C/24 h.

The seeds used were spinel, TiO_2, α-Al_2O_3 and γ-Al_2O_3. When heated to 1100°C, precursor (ii) produced only spinel under definite conditions, though the crystallographic parameters showed some variations.

Precursor (iii) was almost similar in reactivity except when seeded with α-Al_2O_3. Precursor (i) yielded spinel with the same seed, but also MgO when unseeded.

Finally, the mixed alkoxides (precursor iv) did not yield pure spinel under any conditions. It was further observed that isostructural seeding caused a decrease in the temperature of crystallization due to nucleation and epitaxial growth on the seed surface.

Thermal decomposition of mixed and processed salts is another route employed for the synthesis of $MgAl_2O_4$. Messier and Gaza [381] mixed $Mg(NO_3)_2.6H_2O$ and $2Al(NO_3)_3.9H_2O$ and co-melted them in a hood at 230°C. The quenched melt was vacuum dried at 60°–90°C and thermally decomposed in air at 400°C, when the formation of spinel started. The product was ground and calcined at 1000°–1100°C. The spinel thus obtained contained only 10–100 ppm of different impurities, and had a surface area of 190 m^2/g at 400°C, decreasing to $6m^2/g$ at 1100°C.

The versatile combustion synthesis method [226] has also been used for spinel synthesis. The metal nitrates were used (with urea) to prepare the combustion precursors. The single phase spinel particles obtained were 0.1 μm in size, with an average agglomerate size of 5.2 μm and a specific surface area of 21.80 m^2/g.

4.1.10 Titanates and Zirconates

Barium Titanate

Two basic modifications are known for barium titanate, $BaTiO_3$: the high temperature, hexagonal and the low temperature, perovskite-type (130° to ~ 1460°C) structures. Below the Curie temperature (~ 130°C), $BaTiO_3$ assumes a tetragonal structure. Orthorhombic and rhombohedral structures are also obtained on further cooling [382].

The early technique of preparing commercial $BaTiO_3$ powders was via solid state reactions. In this classical method, $BaCO_3$ is mixed intimately with crystalline titania particles and the mixture calcined in air at 1000°–1300°C, and often at ~ 1100°C [383, 384]. The reaction takes place through interdiffusion of cations, and the formation of the titanate. Apart from the decomposition reaction, i.e. conversion of $BaCO_3$ to BaO, the overall reaction involves various intermediate reactions and products, e.g. Ba_2TiO_4, $Ba_6Ti_{17}O_{40}$, $BaTi_4O_8$ etc. [385, 386]. Model experiments have shown the reaction to take place through the following steps :

First step : $BaCO_3 + TiO_2 \rightarrow BaTiO_3 + CO_2$ ... (4.29)

Second step : $BaCO_3 + BaTiO_3 \rightarrow Ba_2TiO_4 + CO_2$ ... (4.30)

Third step : $Ba_2TiO_4 + TiO_2 \rightarrow 2BaTiO_3$ ... (4.31)

The third step is thought to be composed of two sub-steps :

$$Ba_2TiO_4 \rightarrow BaTiO_3 + \text{"BaO"} \quad \text{... (4.32)}$$

$$\text{"BaO"} + TiO_2 \rightarrow BaTiO_3 \quad \text{... (4.33)}$$

The final particle size of $BaTiO_3$ obtained by this solid state reaction is generally relatively coarse (~ 1 μm) due to agglomeration and growth of the TiO_2 particles. The products are also known to be compositionally inhomogeneous [383]. Further, milling of the agglomerates causes pick-up of external impurities (Al, Si, P etc.) and affects the properties of $BaTiO_3$. In spite of the above defects, the process continues to be used industrially. However, the mixed oxide powders thus produced are known to be unsuitable for modern multilayer capacitors [383], where powders free from Pb, Cd, Bi, Li and B with an average size not more than 0.6 μm would be required [387].

One of the well-known commercial techniques of $BaTiO_3$ powder synthesis is via precursor decomposition. The most well-studied precursor is barium titanyl oxalate, $BaTiO(C_2O_4)_2.4H_2O$, commonly abbreviated to BTO. This double oxalate, first reported by Clabaugh et al. [193], was obtained by the slow addition of a mixed aqueous solution of titanium tetrachloride and barium chloride dihydrate into an aqueous solution of oxalic acid at 80°C.

The reaction which can be written as

$$BaCl_2 + TiCl_4 + 2H_2C_2O_4 + 5H_2O \rightarrow$$
$$BaTiO(C_2O_4)_2.4H_2O + 6HCl \quad \text{... (4.34)}$$

yielded a dense crystalline precipitate of the product.

Hydrolysis of titanium, which is an undesirable side reaction in this process, can be restricted to very low levels by keeping the temperature low (at least below 30°C). The pH of the system can be raised for the incorporation of additives, but possibly at the cost of the chemical purity of the precursor [388].

There are controversies regarding the exact intermediate products during the thermal decomposition of BTO. One proposed route is [389] :

$$BaTiO(C_2O_4)_2.4H_2O \xrightarrow{25^\circ-225^\circ C} BaTiO(C_2O_4)_2 + 4H_2O \quad \text{... (4.35)}$$

$$BaTiO(C_2O_4)_2 + 0.5H_2O \xrightarrow{225^\circ-465^\circ C} BaCO_3 + TiO_2 + CO + 2CO_2 \quad \text{... (4.36)}$$

$$BaCO_3 + TiO_2 \xrightarrow{465^\circ-700^\circ C} BaTiO_3 + CO_2 \quad \text{... (4.37)}$$

The stoichiometry and structure of BTO seem to be very much dependent on the conditions of synthesis; this indicates that the "self-adjusted" Ba/Ti mol ratio = 1 in the decomposition product of BTO may not be a matter of course [389–391]. In fact, the process of synthesis of BTO has been modified by various workers [e.g. 390, 391]. Thus, considering the insolubility of Ba- and Ti-oxalates in ethanol and the easy solubility of oxalic acid in the same solvent, Yamamura et al. [391] proposed an alternative method of synthesis of BTO, imparting versatility to the process. Hilton and Frost [383] obtained BTO compounds by the interaction of mixed aqueous solutions of $TiCl_4$ and $Ba(NO_3)_2$ with ethanolic solutions of oxalic acid.

In spite of all these attempts for improving the BTO process, two shortcomings will probably never be removed unless a greater variety of precursor compounds can be synthesized in the same or related systems :

(i) Homogeneous introduction of dopants is not easy by the BTO process.

(ii) The Ba : Ti ratio in BTO is more or less fixed at 1 : 1; so, other compounds of technical interest, e.g. $BaTi_4O_9$, $Ba_2Ti_9O_{12}$, $BaTi_5O_{11}$ cannot be obtained from such a precursor.

A new precursor for $BaTiO_3$ powder, barium titanium glycolate, has been recently reported by Day et al.[392]. This complex salt, $BaTi(C_2H_4O_2)_3.4C_2H_6O_2.H_2O$, has been found to crystallize to $BaTiO_3$ with very minor amounts of $BaCO_3$ at 600°C, and to phase-pure $BaTiO_3$ above 600°C.

Barium titanate is also known to be produced with various morphologies and nanometric size by the hydrothermal technique [388, 393, 394]; Fig. 4.9 provides an example. Various sources of Ba and Ti have been used as reactants under hydrothermal conditions, i.e. elevated temperature (50°–400°C), but often autogenic pressure, in ordinary pressure vessels to obtain fine particles.

Though the hydrothermal technique yielded very fine (20–100 nm) crystalline (commonly cubic) $BaTiO_3$ powder batches with relative ease, the powders were often deficient in Ba-content [388]. One important reason is the presence of residual, unreacted and insoluble Ti-precursors. To make up for this deficiency, excess Ba has been added in the reactant solution (Ba/Ti ratio up to 40!), and the unreacted barium hydroxide washed off with organic (e.g. acetic) acids [393]. With high Ba/Ti ratio, small particles (~ 22 nm) were obtained. Further, crystallization of $BaTiO_3$ has been shown to be easier with higher Ba/Ti ratio, temperature and basicity [394].

An atmospheric pressure "hydrothermal" process has been developed [395] in which excess barium hydroxide (at 70°–110°C) was added continuously in a vigorously agitated slurry of hydrous titania (60°–175°C).

It has been shown [396] that $BaTiO_3$ powder obtained hydrothermally crystallized with a simple cubic perovskite structure, expected only above the ferroelectric Curie point (130°C). Transformation to the tetragonal phase takes place after calcination at 600°C and above. Further, hydrothermal $BaTiO_3$ has been reported to contain considerable amounts of structural OH and H_2O.

Eror and Anderson [397] used the Pechini process [398] for obtaining $BaTiO_3$, in which $BaCO_3$ and Ti-isopropoxide were mixed in stoichiometric proportions in a citric acid-ethylene glycol solution. The solution underwent polymerization and the product was calcined to generate fine particles. More recently, Kumar and Messing [399] proposed a basic pH Pechini process using the same chemicals and obtained pure $BaTiO_3$ above 700°C. Hennings and Mayr [400] prepared a mixed citrate, $BaTi(C_6H_6O_7)_3.6H_2O$, and calcined it at about 600°C to obtain $BaTiO_3$. An elaborate decomposition mechanism was proposed by the authors.

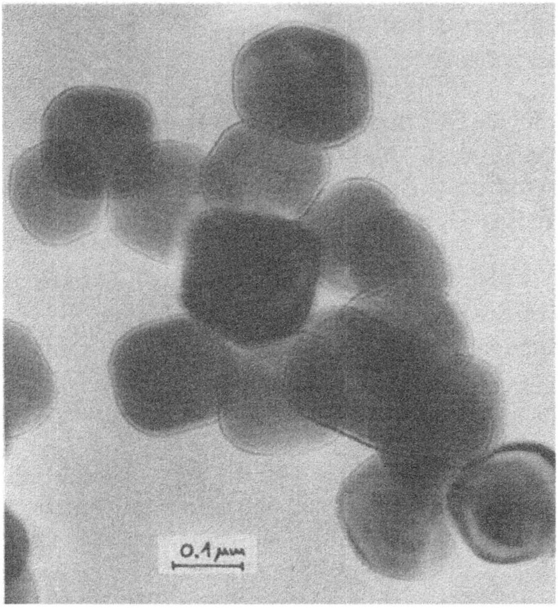

Figure 4.9 Hydrothermally synthesized barium titanate particles, obtained from spray-dried Ti-acetate gel and $Ba(OH)_2$ at pH ≥ 14 and 180°C (kind courtesy : Dr. D. Hennings, Philips Research Laboratories, Aachen, Germany).

Mixed citrates of Ba and Ti were also prepared by Coutures et al. [401] from Ti-isopropoxide and $BaCl_2.2H_2O$ by adding the latter solution into the former at 80°C. With the release of isopropanol, the mixed citrate was spray-pyrolyzed, or dissolved in ethylene glycol to obtain a brown resin after polymerization at 150°C. The products were calcined up to 700°C to obtain 150 nm agglomerates. A recent work by Miot et al. [402] used the same method as above to obtain $BaTiO_3$ of the correct stoichiometry at 700°C.

An elaborate procedure for obtaining fine particles of $BaTiO_3$ through the formation and thermal decomposition of a catechol complex of Ba and Ti has been reported by Ali and Milne [403]; an aromatic diol, $C_6H_4(OH)_2$ was used.

As the first step of the process, catechol was added to dry toluene, and the combination boiled to obtain a solution. A toluene solution of $TiCl_4$ was added dropwise into it under stirring, so as to obtain a dark red precipitate while the heating continued for nearly 1 h. The precipitate $H_2[Ti(cat)_3]$ thus obtained was filtered, washed with hot toluene and dried under reduced pressure.

In the second step, the Ti-complex was added slowly to a water suspension of $BaCO_3$ under boiling conditions. An aqueous solution of $BaTi(cat)_3$ thus obtained was cooled and filtered. The product was freeze-dried and calcined at 600°–1100°C to obtain $BaTiO_3$ (< 2 μm agglomerate size).

Pioneering work on the synthesis of $BaTiO_3$ from alkoxides was reported by Mazdiyasni [161] who demonstrated the preparation of ultrafine (50–150 nm, maximum agglomerate size < 1 μm), high-purity powders by this process. The synthesis occurred via the hydrolysis reaction of refluxed solutions of highly pure barium bis-isopropoxide and titanium tetrakis-tertiary amyloxide dissolved in isopropanol or benzene.

Bernier [404] described a method in which freshly prepared Ba-ethoxide was mixed with Ti-ethoxide and dissolved in ethanol under inert and dried gas flow. Hydrolysis was carried out at –40°C by the addition of an alcoholic solution of water under vigorous stirring. The gel thus obtained, when dried, led to the formation of small flakes. Crystallization took place at 725°C, and the particles thus generated were 0.1–0.2 μm in size, bound together as weak agglomerates.

Kirby [405] refluxed Ba-cyclohexanebutyrate and Ti-2-ethylhexoxide in anhydrous ethanol and added excess water in the solution under continued stirring. Calcination of the product at 800°C yielded only $BaTiO_3$. Complete crystallization of $BaTiO_3$ was also obtained from double isopropoxide solutions as the starting material.

Her et al. [406] generated submicron particles of $BaTiO_3$ by simultaneous introduction of $BaCl_2$, $TiCl_4$ and NaOH solutions at constant flow rates into a reactor containing an NaOH solution at 85°C (forced hydrolysis) in air. Compositions closest to the stoichiometry (0.997–1.007) were obtained with $[BaCl_2]/[TiCl_4]$ = 1.07. Submicron particles of specific surface area 12 m^2/g were obtained.

Strontium Titanate

Strontium titanate, $SrTiO_3$ is known to exist in the cubic perovskite form over wide ranges of temperature (room temperature to melting above 1800°C). Subsidiary structures, e.g. Magneli phases appear when non-stoichiometric compounds are obtained [382].

Gallagher et al. [407] have reported the synthesis of (Ba, Sr) titanyl oxalate precursors by following the method of Clabaugh et al. [193]. Strontium was always required in excess to obtain the desired compositions.

Recently, Bhattacharjee et al. [408] prepared a stoichiometric strontium titanyl oxalate by mixing solutions of $SrCl_2.6H_2O$ or $Sr(NO_3)_2$ and $TiCl_4$ or $TiO(NO_3)_2$, and adding the mixed solution to an aqueous solution of oxalic acid at 25°–80°C under stirring. The oxalate precursor precipitate thus obtained was filtered, washed with water, vacuum dried and finally calcined at 1100°C. Co-precipitation of the products of hydrolysis of titanium with the mixed oxalate was found to be a problem, as it increased the Ti/Sr ratio from the stoichiometric requirements. Two remedies were suggested : (a) precipitation of the oxalate at an elevated temperature and (b) addition of excess oxalic acid in the system.

Tuttle and Voigt [409] describe a citric acid technique for powder preparation. Titanium isopropoxide was slowly added into an aqueous solution of citric acid prepared at 70°C. The hydrolysis product of the alkoxide was obtained as a white precipitate, but redissolved on boiling the mix, as also after adjusting the pH to 5–6. A stable Ti-citrate complex was obtained. This was mixed with a Sr-nitrate solution so as to have a Sr : Ti atomic ratio of 1 : 1. A fine mist of this solution was produced through ultrasonic vibration, and the mist was sprayed into liquid nitrogen or nitrogen-cooled isopentane. The freeze-dried product consisted of spherical particles, which were crystallized to $SrTiO_3$ spheres (up to tens of microns in diameter) at 700°C.

The Pechini method was used for obtaining $SrTiO_3$ particles (Section 3.3.4) by Budd and Payne [160]. Peschke et al. [410] developed a modified Pechini method for $SrTiO_3$ which consisted of the following steps :

(i) Addition of Ti-n-butoxide to ethylene glycol and heating at 75°C to obtain a solution.

(ii) Gradual addition of citric acid to this solution with increasing temperature (85°C).

(iii) Addition of polyacrylic acid or mucic acid (for better chelating action) with continued heating and stirring for dissolution.

(iv) Addition of strontium nitrate and water (for dissolution), with heating at > 100°C to boil off the unnecessary water.

(v) Heating of the viscous solutions at 150°C to obtain the glassy resins as precursors.

The resins were heated in a two-step heating schedule, i.e. 400°–550°C and 550°–700°C for the crystallization of $SrTiO_3$.

In a versatile peroxide route described recently by Pfaff [411], an aqueous solution of $SrCl_2.6H_2O$ and $TiCl_4$ were added rapidly at 10°C under argon atmosphere into 1.5 times in volume of a solution of H_2O_2 and ammonia in water. The molar ratios were: $SrCl_2.6H_2O : TiCl_4 : H_2O_2 : NH_3 = 1 : 1 : 2.5 : 12$. A light yellow precipitate was obtained as a result of the mixing :

$$SrCl_2 + TiCl_4 + H_2O_2 + 6NH_3 + 5H_2O \rightarrow$$
$$SrTiO_2(O_2).3H_2O + 6NH_4Cl \qquad \text{... (4.38)}$$

The peroxo-precursor thus obtained was washed with water, dried and calcined up to 900°C. Well-crystallized $SrTiO_3$ with a surface area of 15 m^2/g was obtained. The same method could be used for various other titanates of strontium.

Strontium titanate of different stoichiometries has also been prepared recently [412] by a simple solution technique. An aqueous solution of $Sr(OH)_2.8H_2O$ was added to $Ti[OCH(CH_3)_2]_4$ under stirring. A cake-like precipitate was obtained by evaporating out the water. The stoichiometric ($SrTiO_3$) precipitate, when dried and calcined at 120° and 700°C, yielded 97–98% cubic $SrTiO_3$ and the rest, anatase. At ~970°C, a single-phase cubic $SrTiO_3$ was obtained. The particles were around 30 nm in size, and had a surface area of 20–90 m^2/g.

Varma et al. [413] employed the spray drying technique for obtaining $SrTiO_3$ powders from sols or dispersed gel particles. The sols were prepared from a titanium isopropoxide solution in isopropanol-acetic acid and an aqueous solution of strontium nitrate. Calcination of the dried product yielded $SrTiO_3$ as almost the single phase at 650°C onwards.

Pyrolysis of oxalate precursors for the synthesis of $SrTiO_3$ (and other titanates) has been reported by Amala Sekar et al. [414]. Strontium titanyl oxalate, $SrTiO(C_2O_4)_2.4H_2O$, oxalic acid dihydrazide and NH_4NO_3 were dissolved in a minimum volume of water in the molar ratio of 1 : 1 : 7 and introduced into a furnace maintained at 350°C. Automatic ignition, preceded by boiling and foaming of the solution, took place with a foam temperature of about 1100°C; titanate particles were obtained within 5 min. The particles were 0.08 μm in size (average agglomerate size ~1 μm) and had a surface area of ~17–28 m^2/g, depending on the conditions of preparation.

Lead Zirconate

A common method for the preparation of lead zirconate, $PbZrO_3$(PZ) powders is by the co-precipitation of the components. In a typical investigation, a mixed solution of $ZrOCl_2.8H_2O$ and $Pb(NO_3)_2$ in the required stoichiometry was added to a 12N ammonia solution under stirring [415]. The mixed hydroxides thus obtained were kept in agitation for 0.25 h, followed by filtering and drying at 120°C/16 h.

PZ powder was obtained after calcining the wet-milled powder at 700°C. The as-prepared precursor was amorphous up to 400°C. PZ dissociated at 1000°C, forming monoclinic ZrO_2.

The citrate method used by Rao et al. [416] involved the addition of citric acid to an aqueous mixed solution of $Pb(NO_3)_2$ and $ZrO(NO_3)_2$. Neutralization of the resulting white coloured thick solution with ammonia produced a clear, transparent solution. Continuous heating of the solution at 300°C gave rise to a powdery material which produced PZ (stable up to 1000°C) at 600°–700°C/1 h. The average size of the powder was 0.1–0.2 μm.

In a sol-gel method, PZ powder has been prepared by dissolving lead acetate and zirconium propoxide in methoxyethanol [417]. Hydrolysis and polycondensation in the solution gave rise to a gel. The gel after calcination at 800°C crystallized to PZ.

Water from a mixed solution of lead nitrate, zirconyl nitrate and urea was abstracted using a large excess of isopropanol under stirring [418]. Decomposition of the precipitated nitrate was carried out at 400°C to form the oxide. Continued heating up to 700°C/6 h led to the formation of pure PZ. The particle size of the powder was 0.4 μm. Evaporation of a mixed solution of $Pb(NO_3)_2$ and $ZrO(NO_3)_2$ [418] also produced PZ at 700°C/6 h. Prismatic particles of cross-section ~ 0.6 μm and length ~ 3.7 μm were obtained by this method.

Lead Titanate

The compound lead titanate, $PbTiO_3$ (PT) forms two polymorphs [419] : (i) the low temperature tetragonal form and (ii) the high temperature cubic form. The transformation occurs at ~ 490°C.

In the commonly used co-precipitation method [420], lead titanate has been prepared by adding a mixed solution of $Pb(NO_3)_2$ and $TiCl_4$ (containing hydrogen peroxide) to an ammonia solution of pH $\cong$ 10 at 40°–45°C The precipitate after washing, followed by drying at 100°C, was calcined at 400°–800°C for zero to as much as 262 h. Fully crystalline tetragonal PT powder of size 100–400 nm could be obtained by this method.

Decomposition of lead titanyl oxalate (PTO) precursor [421] at 600°C is known to produce PT powder. $Pb(NO_3)_2$ and $TiCl_4$ solutions were used as starting materials for the synthesis of this precursor; $TiCl_4$ was converted to $TiO(NO_3)_2$ by precipitating titanium hydroxide with ammonia solution, washing the precipitate with water to remove Cl^- and re-dissolving it in HNO_3. Addition of the mixed solution of $Pb(NO_3)_2$ and $TiO(NO_3)_2$ to an oxalic acid solution (pH < 1) at ambient temperature under vigorous stirring led to the formation of the precursor PTO as a precipitate. The presence of a small amount of PbC_2O_4 along with PTO helped in the formation of stoichiometric $PbTiO_3$ by compensating the loss in Pb during calcination.

The hydrothermal method [422–424] has been based on the treatment of an aqueous solution containing lead and titanium ions in the temperature range of 130°–250°C under pressure. Oxides, hydrous oxides, nitrates, acetates, chlorides, perchlorates or alkoxides have been used as sources of lead and titanium. In a typical process [422], addition of an aqueous solution of lead acetate into titanium isopropoxide in ethanol caused instantaneous hydrolysis of titanium isopropoxide. The entire mass (pH ~ 9.5) was treated hydrothermally between 200° and 300°C at 1.75–9.5 MPa. PbO and titanium butoxide or reactive titanium oxide gels were also used as sources of lead and titanium ions. High-purity, submicron (0.17 μm) PT powder was obtained in this way [423]. When KOH solution was used as the reaction medium [424] at pH 12.5–14.4, PT powder with acicular morphology was formed within 0.5–2 h. The diameter of such particles was < 100 nm and the length, > 1 μm.

Yamaguchi and co-workers [425] obtained a mixed alkoxide by refluxing a mixture of lead isopropoxide and titanium isopropoxide in required proportions for 10 h at 82°C. Hydrolysis of this mixed alkoxide was carried out at 25°C under stirring with distilled water; the temperature of the solution was increased to 80°C. The product was separated by filtration, washed with hot water and dried. The as-prepared powder was amorphous with an average size of 20 nm. At 680°C cubic PT was obtained. The particle size of the powder increased from 200 nm to 1200 nm by increasing the temperature from 500° to 900°C.

Calzada et al. [426] prepared PT powder by the sol-gel method using lead and titanyl nitrate solutions as the source materials. The titanyl nitrate solution was prepared from $TiCl_4$ in the way already described. Co-precipitation of Pb^{2+} and Ti^{4+} as hydroxide from the mixed nitrate solution was effected with NH_4OH. A stable sol was obtained by dispersing the washed precipitate in water and adjusting the pH to 7.0–9.0 with HNO_3. A sol-gel transition was accomplished by evaporating the sol at 60°–70°C; further dehydration gave $PbTiO_3$ gel particles. Phase-pure $PbTiO_3$ crystallized at 500°C.

Lead acetate trihydrate and $Ti(OEt)_4$ or $Ti(O–i–Pr)_4$ or $Ti(O–n–Pr)_4$ have also been used as starting materials for the preparation of precursor powder of PT [427] by the sol-gel method. The method involved the dissolution of lead acetate in 2-methoxyethanol followed by heating at 123°C when a gold coloured solution was obtained. The required amount of titanium alkoxide was added to the acetate solution at the same temperature. Hydrolysis of the diluted solution at –20° to –25°C followed by warming to room temperature caused the formation of a transparent gel. Drying of the gel under vacuum with subsequent grinding and sieving ultimately produced PT-precursor powders.

Lead Zirconate Titanate/Lead Lanthanum Zirconate Titanate

Lead zirconate titanate, Pb(Zr, Ti)O_3 (PZT) is a solid solution of $PbZrO_3$ and $PbTiO_3$ [419]. Depending on the composition and calcination temperature, rhombohedral, tetragonal and cubic forms of PZT are obtained. Lanthanum as a dopant replaces some of the lead ions in the structure, which generates lead lanthanum zirconate titanate, (Pb, La) (Zr, Ti)O_3 (PLZT).

A frequently used technique for the preparation of PZT/PLZT is the co-precipitation method. Co-precipitation of the cations as hydroxides has been accomplished in several ways :

(a) MacNamara [428] used PbO, $ZrOCl_2.8H_2O$ and $TiCl_4$ in 2% HCl as the starting materials. PbO was converted to $Pb(NO_3)_2$ by treatment with HNO_3. The composite nitrate solution was prepared by two different methods :

(i) $ZrOCl_2.8H_2O$ and $TiCl_4$ were separately precipitated as hydrated oxides by adding NH_4OH at room temperature and pH = 7–8 to the chloride solutions. The precipitates were made chloride-free by washing with water and re-dissolved in HNO_3 to form the respective nitrate solutions. The three nitrate solutions of Pb, Zr and Ti were mixed together to obtain the composite nitrate solution.

(ii) $TiCl_4$ in 2% HCl and $ZrOCl_2.8H_2O$ were mixed and cooled to 15°–20°C. Co-precipitation of Zr and Ti was achieved by NH_4OH as described above. The precipitate (made chloride-free) was dissolved in HNO_3 to obtain a mixed nitrate solution of Zr and Ti. Addition of the $Pb(NO_3)_2$ solution to the mixed nitrate solution of Zr and Ti produced the composite nitrate solution.

The composite nitrate solution, either (i) or (ii), was then sprayed on to the surface of a large volume of constantly agitated bath of alkaline ammonium nitrate solution by passing ammonia gas diluted with nitrogen through the bath at 55°–60°C and a pH of 7–8. A mixed hydroxide was precipitated. Calcination of the acetone-washed hydroxide precipitate at 650°C produced high-purity, non-agglomerated PZT powder.

Thomson [170] later prepared PLZT powder following the procedure of MacNamara [428], where a clear solution of the four cations, i.e. Pb, La, Ti and Zr was obtained by adding HNO_3 to a mixed solution of $Pb(NO_3)_2$, $La(NO_3)_3.6H_2O$, $ZrO(NO_3)_2$ and $Ti(OC_3H_7)_4$. The PLZT powder was obtained at 550°C/16 h. The particle size of the powder was ~ 0.5 μm which increased to about 1 μm after calcination at 950°C.

(b) To suppress the hydrolysis of $TiCl_4$ in water, Murata et al. [150] added a mol of H_2O_2 per mol of PLZT to a mixed solution of $Pb(NO_3)_2$, $La(NO_3)_3.6H_2O$, $ZrOCl_2.8H_2O$ and $TiCl_4$. To dissolve the precipitate of insoluble $PbCl_2$ formed during the mixing of the four salt solutions, the mixed solution was heated at 40°C. Co-precipitation was effected by adding this clear solution to an ammonia solution of pH = 8.75–9.25 under constant stirring. The water-washed precipitate, after drying at 100°C, was calcined at 650°C/2 h to obtain fine particles of PLZT as the single phase. The particle size of the powder was 0.1–0.5 μm.

(c) Yoshikawa and associates [429, 430] followed several techniques for the preparation of PZT/PLZT powders using $Pb(NO_3)_2$, $ZrO(NO_3)_2.2H_2O$, $La(NO_3)_3.6H_2O$, $TiCl_4$ or Ti-isopropoxide as the source materials. The $TiCl_4$ solution was converted to a $TiO(NO_3)_2$ solution, according to necessity, as described before. The methods are :

(i) The mixed nitrate solution of Pb, La, Zr and Ti was added to an ammonia solution under stirring at pH ~11 following the reverse strike method [430].

(ii) Hydrogen peroxide was added to the mixed nitrate solution [150] and co-precipitation was allowed as described in c(i).

(iii) The mixed nitrate solution along with H_2O_2 was stirred followed by cooling to 1°C . Co-precipitation was effected by hydrolyzing the solution with ammonia gas introduced to the liquid surface using Ar as the carrier gas [429, 430]. The pH was kept at 10.5–11.0.

The oxyhydroxide precipitates obtained from methods (i), (ii) and (iii) were ultrasonically washed with water followed by acetone and dried at 40°C. The as-dried material was amorphous. The specific surface areas of the products of the above three methods were 84 m^2/g, 140 m^2/g and 135 m^2/g respectively [430]. The as-dried product of method c(iii) consisted of 10 nm sub-units forming spherical agglomerates of about 60 nm size [429]; the sub-units increased to about 20 nm after calcination at 400°–500°C. Crystallization started at about 400°C.

Biggers and Venkataramani [431] hydrolyzed a mixture of zirconium and titanium butoxides under vigorous stirring with a lead acetate solution in water. The resulting hydrolyzed product was dried at 120°C followed by ball milling in ethanol medium. In some experiments PbO was used in place of lead acetate. The precursor material was calcined at different temperatures, e.g. 500°–1000°C for 2–16 h. Single-phase PZT was obtained at 600°C and above.

Duran and Moore [432] synthesized PZT powder following the above procedure [431]. PbO was used as the source of Pb^{2+} and the hydrolysis of the mixed butoxide solution of Zr^{4+} and Ti^{4+} was effected with a water-isopropanol mixture. The as-precipitated powder, dried at ~ 90°C in a highly humid atmosphere, was amorphous up to 400°C and crystallized to a single-phase PZT at 500°C/16 h. The as-precipitated powder showed a surface area of ~ 40 m^2/g which was drastically reduced to 3 m^2/g at 750°C.

Yamamura et al. [433] prepared PLZT powder by the decomposition of an oxalate complex at 800°C. The oxalate complex was obtained when an ethanolic solution of oxalic acid was added to an aqueous mixed nitrate solution of Pb, La, Zr and Ti at 30°C. A part of the oxalate precipitate was washed with a large excess of ethanol. After calcination, specific surface areas of the washed and unwashed precipitates were 9 m^2/g and 3 m^2/g respectively. The average particle size of the washed and unwashed calcined powders was found to be 0.33 μm and 0.64 μm respectively.

However, in the above method, incomplete precipitation of the constituent ions in the solution gave rise to inhomogeneity in chemical composition. The problem was overcome [434] by raising the pH to ~ 10 with NH_4OH before filtration. The oxalate precipitate washed with ethanol was calcined at 1000°C/2 h to obtain chemically homogeneous PLZT powder.

Citrate salts of PLZT components were prepared [435] either by alcohol/freeze drying or gel formation; Pb- and La-oxide, $ZrOCl_2.8H_2O$, Ti-butoxide and citric acid were used as the starting materials. Interestingly, PLZT crystallized at temperatures below 375°C after decomposition of the citrate salt. The specific surface area and the agglomerate size of the calcined powders were about 11 m^2/g and 1 μm respectively.

Bastos and co-workers [436] synthesized citrate precursors of PZT by using $Pb(NO_3)_2$, $ZrOCl_2$, $TiCl_4$ and citric acid as the main ingredients. PZT powder was crystallized at temperatures as low as 500°C from the amorphous material generated by the decomposition of citrate precursors.

Thermal decomposition of metal-EDTA complexes [437] derived from nitrate salt solutions was used for the preparation of PZT powder at a low temperature. The in-situ oxidizing atmosphere generated by the nitrate ions led to the decomposition of organic materials and the resulting heat caused the crystallization of PZT at an unusually low external temperature, i.e. 250°C. The method involved the addition of a mixed metal nitrate solution of Pb, Zr and Ti to an EDTA solution adjusted to pH = 5. The precursor material was obtained by vacuum evaporating the clear metal nitrate solution containing EDTA at 50°–80°C and calcined at 200°–1000°C/1 h in air. The powder obtained at 250°C could produce PZT along with some PbO, and was porous and friable. At 430°C, phase-pure PZT formed. The particle size of the powder calcined at 700°C was 0.1–0.3 μm which increased to 0.5–0.8 μm at 800°C. The chloride-EDTA precursor failed to produce the PZT phase under any condition.

Kutty and Balachandran [438] prepared PZT powder hydrothermally from titania/zirconia mixed gel and PbO in distilled water at 150°–400°C/6-8 h. The Cl^--free mixed gel was obtained by adding NH_4OH to a solution of $TiOCl_2$ and $ZrOCl_2$ as described earlier. For obtaining PLZT powder, freshly ignited La_2O_3 was mixed with PbO. The product was washed with water followed by acetone. Acicular crystals of phase-pure PZT appeared at a minimum of 300°C.

Beal [439] obtained PZT powder hydrothermally in presence of mineralizers like alkali halides and hydroxides in addition to zirconia/titania mixed gel and PbO (10% excess). The mixture was treated hydrothermally at 300°C under a pressure of 7–8 MPa. after cooling to 70°–80°C, the product was finally washed to neutral pH. Low mineralizer concentration has been known to produce rounded or cube-like particles (submicron to several micrometers). Lee et al. [440] showed that with a high concentration (e.g. 10 M) of KOH solution, the temperature of single-phase PZT formation decreased to 100°C/5 h when the starting materials were $Pb(NO_3)_2$, $ZrOCl_2$ and $TiCl_4$. In the absence of KOH, PZT did not form.

Another way of synthesizing PZT hydrothermally was to treat a mixture of aqueous solutions of $Pb(CH_3COO)_2$ and $ZrOCl_2$ and an ethanolic solution of $Ti(OC_4H_9)_4$ along with a KOH solution of different concentrations (1M to 5M) in a teflon-coated autoclave at 100°–220°C/2 h [441]. The excess PbO was washed out before drying the product in air at 100°C. Well-defined PZT crystallized at a temperature slightly higher than 150°C.

Ohba et al. [442] tried to explain the morphology and size of the PZT particles on the basis of the chemistry of the hydrothermal conditions. In this investigation, a mixture of zirconia and titania gels, prepared as described earlier, along with a mixed solution of KOH and $Pb(NO_3)_2$, was charged in a teflon-lined stainless steel autoclave at 150° or 180°C/4 h. The cooled product was washed with water. The amount of PbO and concentration of KOH controlled both the morphology and size of the PZT particles. The particle size decreased with increase in KOH concentration. In fact, the number of nuclei, the degree of supersaturation with respect to PZT and the amount of undissolved PbO controlled the size and morphology of PZT particles. Large euhedral cube-like particles were obtained when the concentration of PbO was 0.3M and that of KOH, less than 2.0 M.

Takigawa et al. [106] followed the spray pyrolysis technique for generating PLZT powder using a mixed solution of lead nitrate, zirconyl nitrate, tetraisopropyl titanate and lanthanum nitrate (prepared by dissolving La_2O_3 in HNO_3 solution) as the starting materials. Mixed solutions containing ethanol were also prepared. Solutions atomized in air, O_2 or N_2 were introduced into the furnace chamber at a constant flow rate. The pyrolysis temperature was varied from 700° to 1000°C. Phase-pure PLZT powder appeared at 900° and 1000°C.

The average agglomerate size of the powder was 2–4 μm. Addition of ethanol had a remarkable effect on the size and particularly the shape of the final product; in presence of 20 vol% ethanol, small, highly spherical particles of PLZT with uniform size and chemical homogeneity were obtained.

In another popular technique, PZT/PLZT powders have been prepared by the spray or freeze drying of (i) a mixed nitrate solution [170], and (ii) a suspension of the mixed cation hydroxides [443] to generate powders of homogeneous composition. In method (i), $ZrOCl_2.8H_2O$ and Ti-isopropoxide used as the source materials of Zr and Ti were converted to the respective nitrate solutions following methods described earlier. The composite nitrate solution was then subjected to spray drying for preparing compositionally homogeneous powders. Spray-dried powders were calcined to decompose the nitrate. Phase-pure PZT crystallized at 550°C.

Freeze drying [170] of the same solution in a cold (–70°C) hexane bath produced free-flowing, spherical PLZT powders. In actual practice, the solution was sprayed into the hexane bath. Water from the solid spheres of the salt was removed by sublimation in the freeze dryer. The particle size of the spray-dried and freeze-dried powders obtained by Thomson [170] was found to be 0.1–0.2 μm at 550°C which increased to ~ 1 μm at 950°C.

In method (ii) above, a suspension of hydrated Pb/La/Zr/Ti-oxide in a 60 : 40 isopropanol : water mixture was prepared using $Pb(NO_3)_2$, $La(NO_3)_3.6H_2O$, $ZrOCl_2.8H_2O$ and $TiCl_4$ as the starting materials. The co-precipitation reaction was performed following the procedure originally described by Murata et al. [150]. The Cl^--free precipitate was used to prepare a suspension in 60 : 40 isopropanol : water for spray drying. The spray-dried powders were calcined at 400°–700°C.

The product consisted of solid particles of spherical morphology with a relatively wide size distribution (0.1–3.0 μm). The dried powder was X-ray amorphous, but crystallized to single-phase, compositionally homogeneous PLZT at 600°–700°C with the retention of sphericity. Presence of the organic solvent improved the sphericity of the spray-dried powders which showed a specific surface area of about 110–120 m^2/g [443]. The value decreased to 10 m^2/g for the powder calcined at 900°C. The same suspension, when subjected to freeze drying in a liquid nitrogen bath, produced agglomerated particles, heat-treated to generate PLZT.

Hydrolysis of the mixed alkoxide with water is another method for preparing submicron PZT/PLZT powders. Brown and Mazdiyasni [444] used a mixture of lead isoamyloxide and lanthanum isopropoxide along with zirconium and titanium tertiary amyloxide in a mutual solvent, i.e. isoamyl alcohol and hydrolyzed the solution by adding it slowly to triply-distilled deionized water to obtain PLZT powder. The hydroxide powder was washed with water and isopropanol, followed by drying at 60°C under vacuum. The as-precipitated powder was amorphous to X-ray up to 500°C and cubic in morphology. On further increase in temperature from 600° to 750°C, crystalline, cubic PLZT was obtained. Extensive agglomeration occurred after calcining the submicron powder either at 500°C for longer time or at higher temperatures.

Monodisperse PZT precursor powder was prepared [165] by refluxing a mixture of lead isopropoxide, zirconium tetrabutoxide and titanium ethoxide in ethanol at 74°C/48 h in a nitrogen atmosphere, followed by hydrolysis of the refluxed alkoxide with water at room temperature. The particle size, morphology and the state of agglomeration varied with the pH and water concentration in the system. Monodisperse spherical particles of PZT were obtained when hydrolysis was carried out at pH = 3 or 7 keeping other parameters unchanged; the particles had a mean diameter of 0.4 μm and 0.7 μm respectively. The as-prepared powder was amorphous with a specific surface area of 6.0 m^2/g and at 600°C, crystalline, homogeneous tetragonal PZT was obtained.

Hirashima et al. [445] demonstrated the formation of PZT powder by hydrolyzing a mixed alkoxide of $Pb(iso\text{-}C_3H_7O)_2$, $Zr(n\text{-}C_4H_9O)_4$ and $Ti(iso\text{-}C_3H_7O)_4$ at 25°–76°C in ethanol or at 95°C in n-propanol with a large excess of water. Hydrolysis in ethanol medium at temperatures below 76°C produced spherical particles with a mean diameter of about 0.34 μm. The powder crystallized to PZT at 400°C/72 h. On the otherhand, hydrolysis in n-propanol at 95°C developed small, irregular-shaped particles.

PZT powder was also obtained by the non-alkoxide sol-gel method [446] which involved the mixing of $Pb(NO_3)_2$, $ZrOCl_2$ and $TiCl_4$ solutions under stirring; NH_4NO_3 was added to the mixed solution. Co-precipitation of the mixed solution was carried out with a 12N ammonia solution. A stable sol was obtained by dispersing the thoroughly washed complex hydroxide in water. Addition of a dehydrating agent, e.g. 2-ethylhexanol along with a surface active agent formed a gel. Calcination at 600°C/1 h or 800°C/1 h produced PZT as a single-phase material. The average particle size of the powder was less than 5.5 μm.

Tomandl et al. [447] prepared sol-gel PZT powder containing 1 mol% neodymium oxide as the sintering additive using lead acetate, neodymium oxide, Zr-n-propoxide and Ti-ethoxide as the starting materials. The method involved the dissolution of lead acetate in methanol to which an acetic acid solution of neodymium oxide was added. Dropwise addition of a mixture of Zr-n-propoxide and Ti-ethoxide to the above solution was carried out under vigorous stirring. After hydrolysis with water, the temperature was increased to 95°C when the viscosity of the resulting sol went on increasing by polymerization and evaporation; the viscous sol formed a gel. Calcination of the gel material at 600°C/10 h formed single-phase PZT which after milling for 0.5 h produced PZT powder.

Following the method adopted by Gurkovich and Blum [448] for the preparation of lead titanate gel, Zhuang et al. [449] synthesized PZT powder. For this purpose, lead acetate was dissolved in 2-methoxyethanol followed by boiling the solution at 125°C (the boiling point of 2-methoxyethanol) for dehydrating the liquid. After cooling the solution to 75°C, it was added to a solution of titanium isopropoxide and zirconium-n-propoxide in 2-methoxyethanol followed by boiling at 125°C. Hydrolysis was carried out by adding de-ionized water to the cooled mass at –25°C in a liquid nitrogen/2-propanol bath. Gel formation occurred at room temperature; the gel was dried at 100°C. At 600°C/1 h, PZT crystallized as the single phase.

Ostertag et al. [450] followed three different methods for synthesizing sol-gel PZT powder. The methods are :

(i) To a mixture of Zr-isopropoxide and Ti-ethoxide containing a complexing agent, e.g. acetic acid or acetylacetone, a methanolic solution of lead acetate was added under stirring. Addition of water produced a gel. Removal of the solvent at 110°C caused the formation of a precursor powder.

(ii) A large amount of acetone was added to the sol in (i) above. Instantaneous formation of a white powder of size 100 nm occurred; this was washed with acetone.

(iii) Dilute HNO_3 and excess water were used to hydrolyze an ethanolic solution of Zr-isopropoxide and Ti-ethoxide. The solvents were distilled out; the residue was dispersed in water. The process was repeated for complete removal of the solvent. Addition of a $Pb(NO_3)_2$ solution to the acidic Zr–Ti hydroxide sol produced a composite sol of Pb, Zr and Ti. An emulsion of this sol was prepared under vigorous stirring in petroleum ether containing an emulsifying agent. The precursor powder was obtained by passing NH_3 gas through the emulsion. The powder, after washing with ethanol, was dried.

The precursor materials of methods (i), (ii) and (iii) were calcined at 650°C; phase-pure PZT powder with crystallite size < 300 nm was obtained.

4.1.11 Ferrites

The ferrite family of compounds is divided into three groups [451] : (i) spinel ferrites (both normal and inverse types), (ii) hexagonal ferrites and (iii) garnet ferrites.

In the conventional method of the preparation of ferrite powders, the individual oxides and/or carbonates of the metal ions are mixed together, calcined and ground to fine powders. Although the method is cost-effective, the resulting powder suffers from lack of purity and compositional homogeneity; the particles are relatively coarse with a wide size distribution. For the synthesis of high-purity, chemically homogeneous submicron ferrite powders, various chemical methods are currently followed on a parallel basis.

In the co-precipitation technique, (Ni, Zn)-[452] and (Mn, Zn)-[453] ferrites, barium hexaferrites [147] etc. were prepared from metal nitrates, sulphates or chlorides as the starting materials. Generally, NaOH, NH_4OH, a mixture of NaOH and Na_2CO_3 or NH_4OH-$(NH)_2CO_3$ was used as the precipitant. The nature of the precipitate depended on the pH and temperature of precipitation. The precipitation could be carried out either at ambient temperature or at 50°–95°C. Single phase, chemically homogeneous ferrite particles were obtained after calcination of the thoroughly washed hydrous precipitates at 700°–950°C [147]. The particle size of the co-precipitated material was ~ 4 nm [453], while that of the calcined powder varied from 0.1 to 0.6 μm [147]. High surface areas (> 100 m^2/g) of the oxide particles were also noted [452]. Precipitation was reported to be complete at pH > 11; below this pH value the cations remained in the solution.

Addition of a 5–10% solution of 8-hydroxyquinoline [454] in acetone to a mixed salt solution in presence of tartaric acid and ammonia could produce ferrite powders. Under such conditions, certain elements, e.g. Fe, Cd, Co, Mn, Mg, Cu, Ni and Zn were selectively precipitated while Na, K, Ba, SO_3^{2-}, PO_4^{2-} etc. remained in the filtrate. Calcination of the precipitates, i.e. hydroxyquinolinates, at about 400°C removed the 8-hydroxyquinoline, resulting in the formation of the mixed powders.

Gallagher and co-workers [455] prepared nickel ferrite by the decomposition of the co-precipitated crystalline oxalate precursor, $Ni_{0.2}Fe_{0.8}C_2O_4.2H_2O$, at a temperature of ~ 500°C. The precursor was obtained by the addition of a hot solution of ammonium oxalate to a heated mixed sulphate solution of Ni^{2+} and Fe^{2+}; the resulting slurry was kept at 75°C/1 h. The precipitate was filtered, washed and dried.

The hydrothermal process produced highly crystalline ferrite powders at low temperatures. For the preparation of barium hexaferrite $BaFe_{12}O_{19}$, nitrate solutions of Ba^{2+} and Fe^{3+} and an excess of NaOH solution were heated to 200°–300°C in a stainless steel autoclave followed by cooling to room temperature [456]. The experiment was conducted in nitrogen atmosphere. Ferrite powders with the compositions $BaFe_{12-2x}Ti_xCo_xO_{19}$ (x <1.2) were formed when a nitrate solution of Co^{2+}, and $TiCl_4$ were mixed with the above solution. The particle size of the ferrite powders was 0.1–0.3 μm.

Ataie and co-workers [457] investigated the role of inorganic and organic bases during the formation of barium hexaferrite by hydrothermal processing. Platelike particles of barium hexaferrite were formed when solutions containing $Ba(NO_3)_2$ and $Fe(NO_3)_3.9H_2O$ along with different base species like NaOH, KOH, NH_4OH and $(C_2H_5)_4NOH$ were treated hydrothermally at 220°C/5 h. Both the size and thickness of the synthesized particles varied from 1.2 to 0.8 μm and from 0.28 to 0.15 μm respectively in presence of inorganic bases. When the organic base, $(C_2H_5)_4NOH$ was used, the thickness decreased to about 0.06 μm.

Agglomerate-free ferrite particles with controlled stoichiometry and narrow size distribution were prepared by the spray pyrolysis technique. For the preparation of barium hexaferrite [458–460] and strontium hexaferrite [461] particles, an aerosol stream of the mixed nitrate solution was passed through a tube furnace at 450°–1200°C. Pyrolysis of the droplets occurred and ferrite particles were collected in a filter bag. Single-phase ferrite powders were obtained at 900°–1200°C; the powders collected at ~800°C contained Fe_2O_3 as the second phase. The size of the particles varied in the range 0.1–1.0 μm depending on the calcination temperature (800°–1100°C) and the crystallite size was confined to < 100 nm.

Matsumoto et al. [462] generated fine particles of YIG single crystal by the mist pyrolysis of a mixed aqueous solution containing $Y(NO_3)_3.6H_2O$ and $Fe(NO_3)_3.9H_2O$. The mist generated by an ultrasonic nebulizer was passed into a reaction tube at 800°C under a constant flow of N_2. Decomposition of the mist occurred at 800°C. The as-prepared material was amorphous and spherical in morphology; at 1000°C, YIG particles with an average diameter of 0.25 μm were obtained.

Kladnig and Zenger [463] synthesized (Mn, Zn)-ferrites following the Ruthner spray roasting technology [176]. The method involved the spraying of a mixed nitrate solution of Mn, Zn and Fe directly into a furnace heated at a temperature higher than the decomposition temperature of the salts (300°–600°C). This co-spray-roasting method has been used for the preparation of (Mn, Zn)-ferrite powders on an industrial scale. The specific surface area of the ferrite powder was 68 m^2/g due to high internal surface; the average particle size was within 10 μm. Cation impurities in the powders were found to be in the range of 100–700 ppm.

Delau [464] prepared (Ni, Zn)-ferrite powders by the spray drying process using the corresponding sulphate salts. The method involved the atomization of a mixed sulphate solution of Ni, Zn and Fe into droplets of 10–20 μm size by a centrifugal disk atomizer, followed by drying in a stream of hot air. The spray-dried precursor salt was calcined at 800°–1000°C and converted into ferrite powders. The oxide powders consisted of small particles (~ 0.2 μm) as agglomerates which were ground to submicron particles.

In the liquid drying process [465, 212, 466], a homogeneous salt solution of the ferrite materials [212, 466] was atomized by compressed air in highly agitated hygroscopic liquids like acetone or a mixture of acetone and NH_4OH solution in 25 : 1 volume ratio. The dehydrated powder was decomposed at 400°–900°C in air or nitrogen atmosphere. Ferrite particles of spherical morphology varied in size in the range 0.065–0.45 μm depending on the calcination temperature.

Fine ferrite powders were also prepared by the hydrolysis of alcoholic mixtures of metal-organic compounds. Examples are $CoFe_2O_4$ [467] and doped barium hexaferrites [e.g. 468]. A mixed solution of the ferrite constituents was prepared by dissolving calculated quantities of the metal acetylacetonates and alkoxides in anhydrous ethanol followed by heating at 60°C. The hydrolysis reaction was allowed to occur at 60°C with ammonia solution diluted with CO_2-free deionized water at a pH of about 11. The resulting suspension was refluxed for several hours in N_2 atmosphere. The precursor powder was collected by centrifugation and dried in a microwave oven. The as-precipitated hexaferrites were amorphous, while $CoFe_2O_4$ was crystalline. Calcination of the amorphous precursor at 800°–1000°C generated ferrite particles of submicron size. The as-precipitated crystalline $CoFe_2O_4$ ferrite powders had a particle size of ~5 nm, which increased to about 10 nm after heating at 450°C.

Ultrafine particles of amorphous YIG were synthesized by Taketomi et al. [469] by the hydrolysis of a mixture of iron ethoxide in ethanol and yttrium butoxide in xylene. The particle size of the amorphous YIG powder was about 9 nm. Amorphous to crystalline phase transformation occurred at about 700°C.

Brahma et al. [470] synthesized hexagonal barium ferrite powder by the sol-gel method. Anhydrous $FeCl_3$, $Ba(CH_3COO)_2.H_2O$ and ethanol were used as the starting materials. An acetic acid solution of barium acetate was mixed with an ethanolic solution of $FeCl_3$ (prepared by refluxing at 78°C/1 h). The resulting mixed solution at pH = 3 was allowed to gel. The 350°C dried gel after crushing and pelletization was calcined at 900°–1000°C/1–2 h which led to the formation of ferrite along with some unidentified materials.

Ultrafine particles of $SrFe_{12}O_{19}$ were recently prepared by Choy et al. [471] by the citrate gel method. A bulk amorphous gel was obtained by heating a clear solution of $Sr(NO_3)_2$, $Fe(NO_3)_3.9H_2O$ and citric acid at pH = 6.5 (adjusted with NH_4OH). Thermal decomposition of the precursor citrate gel at a considerably low temperature, e.g. 850°C/2 h produced strontium ferrite powder of ~50 nm size.

Instead of forming a bulk gel as mentioned above, a mixed metal complex precursor was obtained by various workers by evaporating a complex solution with citric, maleic, tartaric, glycollic or lactic acid of the metal concerned at low temperature, e.g. 40°–100°C in air [472–474]. Thermal decomposition of the complex precursor led to a gradual transformation of an amorphous material to a crystalline product. The transformation temperature varied from 600° to 1100°C. Chemically homogeneous submicrometer-sized particles of ferrites could be generated by the direct decomposition of the complexes. The particle size varied with the pyrolysis conditions. All types of ferrites, e.g. (Mn, Zn)-ferrite, barium hexaferrite, Bi-substituted YIG etc. could be prepared by this method.

Based on the fact that metal salts can form complexes with hydrazine hydrate and related compounds, ultrafine particles of ferrites were prepared at low temperature using these materials. In one of the methods [475], (Ni, Zn)-ferrite was prepared by the co-precipitation of the mixed metal nitrate solution of Ni, Zn and Fe with hydrazine hydrate, $N_2H_4.H_2O$ and KOH followed by ageing at 90°C/1 h in a closed pyrex glass tube. The product after washing and drying at 50°C was further heated to 150°C. Depending upon the conditions of preparation, the crystallites varied in size from 3 to 30 nm. Recently, Ueda et al. [476] followed the above-mentioned procedure during the synthesis of crystalline Zn-ferrite powder using $ZnCl_2$ and $FeCl_2$, and $N_2H_4.H_2O$ as the source materials. The mixed solution of $ZnCl_2$, $FeCl_2$ and $N_2H_4.H_2O$ was aged at 30°–60°C. Poorly crystalline material was obtained.

In another method [477], ultrafine ferrite particles were obtained both by the combustion in air and decomposition in a melt of the precursors hydrazinium metal hydrazinecarboxylate hydrates, where the metals were Mg, Mn, Fe, Co, Ni and Zn. During the decomposition of the precursors at low temperature (75°–200°C) in air, large amounts of gases evolved. Once the precursors were ignited, the decomposition was autocatalytic, as in various other systems mentioned in this Chapter. In the melt process, a mixture of $NaNO_3$: KNO_3 (1 : 1) with a eutectic temperature of 220°C, which was just above the decomposition temperature of the precursors, was used. During the addition of the precursor to the melt at 240°C, the former decomposed immediately, producing ultrafine particles of ferrites; the products were collected in cold water. The particle size of the ferrites thus obtained was 6–22 nm with specific surface area values of 100–140 m^2/g.

Single phase spinel ferrites, e.g. (Mg, Mn)-ferrites of average particle size 3.5 μm and specific surface area of 50 m^2/g were prepared by the decomposition of the precursor, $(N_2H_5)_3\ Mg_{1-x}Mn_xFe_2(N_2H_3COO)_9.3H_2O$ (x = 0.50 and 0.62) at 250°C in O_2 atmosphere [478]. Voluminous fine powders were obtained.

Using the glass-ceramic method, which is not a very common route for powder preparation, strain-free crystalline particles of ferrites, e.g. $BaFe_{12}O_{19}$ [479] and $SrFe_{12}O_{19}$ [480] with the desired particle size and shape were prepared by controlled heat-treatment of suitable glasses. The method did not require milling of the synthesized particles as the crystalline product was recovered by leaching of the glassy phase by a suitable acid. No impurity phases were found to form in the process. The particles thus obtained had the morphology of hexagonal or rectangular platelets and were 0.2 to 0.5 μm in size.

Ferrite powders with shape anisotropy (plate-like or needle-like powders) are known to be prepared by the molten salt method, employed only for a limited number of ceramic powders. The method involved the heating of a mixture of reactant oxides or carbonates in required stoichiometry in a furnace along with salts, e.g. NaCl, KCl, Li_2SO_4, Na_2SO_4 etc. above their melting points. Usually mixed salts (flux), e.g. Na_2SO_4–Li_2SO_4 or NaCl–KCl [481] or Na_2SO_4–K_2SO_4 [482] were used to lower the melting temperature. The furnace temperature was kept in the range of 700°–1000°C for a desired time period followed by cooling. The reactant oxide or carbonate mixtures were dissolved in the molten salt. After cooling, the product was thoroughly washed and dried. Depending on the conditions of heat-treatment, homogeneous powders in the size range of 0.2–2.0 μm were produced. The morphology of the particles has been found to change after prolonged heating in the molten salt.

In a modified flux system [483], Co^{2+}- Ti^{4+}- substituted barium hexaferrites with homogeneous composition were prepared using only 30 wt% KCl at 950°C/5 h or 1000°C/1 h when the starting materials were oxides or carbonates. The particle size was in the range 0.2–0.4 μm. This modified flux method is considered for the bulk production of ferrite powders with controlled magnetic properties.

4.1.12 Other Oxide Powders

Lead Magnesium Niobate

Among the lead-based relaxor compounds noted for their high dielectric constant, $Pb(Mg_{1/3}Nb_{2/3})O_3$ (in short, PMN) with a perovskite structure is the best known. As a result, various routes have been employed to obtain powders of this compound; a detailed account is given in a recent review [484]. The search for effective synthetic routes is motivated by the fact that (a) phase-pure perovskite PMN is difficult to obtain, as it is often accompanied by a pyrochlore type lead niobate phase, and (b) the temperature of powder synthesis (and subsequent sintering) must be kept low so as to avoid loss of Pb through vaporization (PMN is known to lose 0.5 wt% PbO per hour at 900°C [485])

To take care of the above problem, several strategies have been evolved in addition to the usual route, e.g.

(i) Pre-reaction of Mg- and Nb-chemicals and formation of $MgNb_2O_6$ to ensure that the Mg-free pyrochlore phase does not crystallize (the "columbite" method).

(ii) Pre-reaction of Pb- and Nb-chemicals and formation of $Pb_3Nb_2O_8$ followed by reaction with MgO (the "wolframite" method).

(iii) Addition of excess Pb in the starting compositions.

The above problems and approaches are discussed below in some more detail.

Swartz and Shrout [486], from their studies on the solid state reactions of PbO, MgO and Nb_2O_5, came to the conclusion that a large amount of the undesirable pyrochlore would accompany PMN in the reaction product even up to 1000°C, and that residual Pb and Mg oxides would always be obtained as minor phases, indicating incomplete reactión. The reaction sequence suggested by them was :

$$PbO + Nb_2O_5 + MgO \xrightarrow{25°-650°C} \text{No reaction}$$

$$PbO + Nb_2O_5 + (MgO) \xrightarrow{700°-800°C} \text{Cubic pyrochlore}$$

$$\text{Pyrochlore} + PbO + MgO \xrightarrow{850°-900°C} \text{Perovskite + pyrochlore}$$

Addition of excess MgO (5 mol%) led to the formation of only 80–83% perovskite at 870°–1000°C/4 h. The difficulty of the solid-solid reactions among the constituent oxides to obtain the necessary phase purity has been well-documented [487–489].

Considering the inadequacy of the solid state process, Swartz and Shrout [486] resorted to the pre-reaction of MgO and Nb_2O_5 to obtain columbite-type $MgNb_2O_6$ (mentioned earlier) at 1000°C. When PbO and 2% excess MgO were mixed with the Mg-niobate, pellets fired at 800°C/4 h yielded approximately 100% perovskite. In a later work, Swartz et al. [490] readdressed the question of addition of MgO and PbO and showed that though the latter helped to obtain stoichiometric PMN, the electrical properties showed deterioration. The formation of $MgNb_2O_6$ by solid state reactions has been recently examined by Saha et al. [491]

Guha and Anderson [492] looked at the problem from a different angle: they pre-reacted PbO and Nb_2O_5 to obtain $Pb_3Nb_2O_8$ by solid-solid reaction at 820°C. This product, mixed with MgO (1 wt% in excess), was pressed into pellets, crushed and re-pelletized during firing at 900°C/4 h to achieve complete conversion to PMN and avoidance of loss of Pb.

Lejeune and Boilot [488] pointed out that volatilization of PbO led to the formation of the pyrochlore phase, and recommended the addition of excess PbO. The excess PbO (6–8 wt%) is therefore sometimes added after the calcination of the powders. Though this addition reduced pyrochlore formation, lowered the firing temperature of the ceramic and increased its density, the lead oxide segregated at the grain boundaries and reduced the dielectric constant; further, it is known that the presence of pyrochlore phase does not affect the dielectric properties of PMN so long as the purity is very high (99.999%) [484].

Different solution routes have been used for the synthesis of PMN, some examples of which are discussed below.

Watanabe et al. [493] used the co-precipitation technique for the synthesis, in which the reactants were solutions of lead nitrate, magnesium hydroxide and a hydrated oxide of niobium. The precipitates, washed and dried, were calcined at

800°C, followed by a heating (under oxygen atmosphere) at 900°C/2 h. PMN thus obtained consisted of 0.3 μm particles.

A sol-gel method was evolved by Ravindranathan et al. [494] for PMN. Lead acetate was dissolved in methoxyethanol at 70°/3 h, and the solution refluxed at 125°C/12 h under stirring. After the elimination of water under the flow of argon, the solution was cooled. A mixture of Mg- and Nb-ethoxide was added to the Pb-solution, and the resulting solution was heated to 110°C/12 h. Water was added dropwise into this complex solution to obtain a gel within 24 h. After room temperature drying and heating the gel at 320°C, an amorphous powder was obtained. The results of calcination can be summarized in the following way :

$$\text{Amorphous} \xrightarrow{320^\circ-650^\circ C} \text{Pyrochlore} \xrightarrow{700^\circ-750^\circ C}$$

$$\text{Pyrochlore + Perovskite} \xrightarrow{775^\circ C} \text{Perovskite} + < 2\% \text{ Pyrochlore.}$$

A complex alkoxide of Pb, Mg and Nb was obtained by Fukui et al. [495] by preparing a solution of lead acetate and niobium ethoxide by refluxing them in ethanol/cyclohexane and adding Mg flakes and cyclohexane in this solution. After addition of acetone into the complex alkoxide that was synthesized in the process (acetone : ethanol : cyclohexane = 1 : 1 : 1), the latter was hydrolyzed with ammonia water when precipitation occurred. The suspension was stirred at 80°C/24 h. An amorphous powder with a particle size of ~ 0.3 μm was obtained. Unlike in other similar investigations, this powder crystallized at only 300°C to > 70% perovskite. However, even after heating at 900°C or above, about 10% of the pyrochlore phase still remained unconverted.

An all-alkoxide method was used by Chaput et al. [496] in which lead butoxide and a double ethoxide of Mg and Nb were hydrolyzed to obtain a gel. At 700°C, a pure perovskite phase was obtained in fine powder form.

Anderson et al. [485] employed the Pechini or citrate method for powder preparation in which a variety of precursors were used. The starting materials (carbonates, hydroxides, isopropoxides and nitrates) were mixed with a solution of citric acid and ethylene glycol to prepare a mixed solution; the temperature was kept at 90°C to ensure complete dissolution. For PMN, Pb-carbonate, Mg-carbonate and Nb-ethoxide were the selected starting materials. The mixed solution was evaporated until a polymerized mass was obtained. This was heated to 400°C to drive out the organics. The brittle mass of black colour thus obtained was ground, sieved and calcined at different temperatures for 3 h. The product consisted mostly of the perovskite phase; at 700°C, PMN was the major phase and at 900°C, apparently only PMN was obtained. Particles of < 50 nm size could be obtained up to only 800°C. Excess MgO favoured the crystallization of the perovskite phase, but also raised the sintering temperature.

The citrate route was also used by Choy et al. [497], in which lead and magnesium nitrates and niobium chloride were used. The obtained gel was prefired at 500°C and rapidly heated to 900°C. The rapid firing (thermal shock) has been known to assist avoidance of the undesired pyrochlore phase [484].

Gupta and Kulkarni [498] recently used a "partial oxalate" process where a magnesium niobate was first prepared, and a methanolic dispersion of the niobate sprinkled into a 2 M oxalic acid solution. This followed a dropwise addition of a lead nitrate solution to form Pb-oxalate, which apparently coated the Mg-niobate particles. The product was presintered at 800°C, ground and sintered at higher temperatures.

Y- and Bi-based Cuprate Superconductors

Since the discovery of superconductivity in cuprate ceramics, a large number of new superconducting (high T_c) compounds have been synthesized and the extents of substitutional solid solution possible in them studied in detail. In a recent review, Rao et al. [499] discuss 29 cuprate superconductors. Most of the well-known solid state and (especially) solution routes have been employed for obtaining superconducting compounds. In this Section, only some examples are provided from the vast literature on the synthesis of the superconducting cuprate $YBa_2Cu_3O_7$ (nominal composition), also referred to as YBC, as also from the bismuth-based compounds [499]. Generally those examples have been considered where the precursors have *not been* packed and sintered before crystallization, but first transformed to amorphous or crystalline particles.

One of the early accounts of solid-solid reactions leading to YBC was presented by Chen et al. [500] as a recipé. Dried Y_2O_3, $BaCO_3$ and CuO were used as the reactants which were heated in air at 900°C/9 h after intensive grinding and mixing. Flowing oxygen during the heating was recommended. The charge was removed from the furnace at 400°C and pulverized. The whole heating process was repeated two more times to complete the reaction.

Spann et al. [501] also used Y_2O_3, $BaCO_3$ and CuO as the starting materials for the synthesis ofYBC particles. After dry and wet milling, the mixture in specified ratios was calcined according to the following "single step" schedule for a total of 22 h :

(i) Heating to 950°C under vacuum.
(ii) Flow of pure oxygen during a hold of 4 h.
(iii) Cooling to 250°C before taking out the product to the ambient.

The product was found to be essentially YBC, with minor unknown phases; however, no $BaCO_3$ was observed. This was particularly encouraging, as $BaCO_3$ generally resists easy decomposition during calcination of solid mixtures [502].

The vacuum treatment was considered to be an important aid for the promotion of the reactions. The particle size of the product was 0.25–0.30 μm.

Calcination in air of a specially prepared reactive precursor was reported by Tartaj et al. [503] for obtaining single-phase YBC. The starting materials were $BaCO_3$ in an aqueous suspension, and Cu- and Y-nitrates dissolved separately in water and converted to a mixed Cu/Y solution under stirring. The solution was added to the suspension and a KOH solution added to it for precipitation. The obtained grey-black slurry was filtered, washed, milled, re-washed and finally de-agglomerated in a disc turbine (2000 rpm). This precursor was calcined at 900°C/4 h in air, slowly cooled to room temperature and de-agglomerated under harsh conditions. Porous, soft agglomerates of pure YBC thus obtained had 40–60 μm particles which broke down to < 10 μm after milling. A drawback of the process was the presence of 0.005 wt% K_2O.

The oxalate method of synthesis of YBC has been examined by many investigators [499]. The general procedure involved the addition of an oxalic acid solution in an aqueous nitrate solution containing Y, Ba and Cu; the pH was adjusted at 7.5. The slurry thus formed was digested, filtered and dried. Finally, the product was calcined at 780°C in air for several days, and oxygenated later at 450°C to obtain fine particles of YBC which often had undesirable stoichiometry. The problems associated with this method have been discussed by Rao et al. [499].

In an oxalate method [504] of synthesis of Bi-Pb-Sr-Ca-Cu-O powders, an aqueous chloride solution (30 mol %) of the cations was mixed with an ethanolic solution of oxalic acid. The pH of the mixed solution was brought to 4 by the addition of an ammonia solution under vigorous stirring. The obtained precipitate was filtered, dried and calcined at 750°C/4 h. The product of calcination had a cationic relationship of Bi : Pb : Sr : Ca : Cu = 0.8 : 0.2 : 1.0 : 1.1 : 2.0. The substance was a poorly crystalline mixture of high T_c and low T_c phases. Crystallinity improved after calcination at 835°C for up to 305 h, but a phase-pure, high T_c substance was not obtained.

Some objections have been raised [505] against the conventional oxalate method :

(i) The ammonium oxalate formed during pH adjustment helps the formation of soluble amine complexes (e.g. of copper).

(ii) A very precise control of pH is required.

(iii) Rare earth ions can form undesirable double salts with alkaline earth ions, thereby hindering the precipitation of the individual oxalates in the right stoichiometry.

These authors, therefore, used nitrate solutions of the different cations and reacted them with (ethanolic solutions of) triethyl ammonium oxalate (TEO), sometimes in combination with triethylamine (TEA). The advantages were that (a) TEO had a high degree of solubility in the ethanol-water solution, and (b) TEA did not form a soluble complex with copper.

The citrate gel method has been used by several workers for the synthesis of superconducting compounds. An early and simple work [506] consisted of mixing nitrate solutions of Y, Ba and Cu with citric acid as the first step. When the water was evaporated at 75°C, a white precipitate of Ba-nitrate was obtained, which redissolved on addition of excess NH_4OH (the pH came to ~ 6). The viscous solution did not yield any more precipitates on continued evaporation. When heated in vacuum at 85°C, an amorphous, glasslike dark blue solid was obtained. This solid yielded YBC with a trace of $BaCO_3$ at 900°C, but only YBC at 950°C; the white precipitate mentioned above also crystallized into phase-pure YBC after vacuum drying and heating at 950°C.

In a detailed work involving citrate formation [507], a Ba-EDTA complex was prepared by slowly adding $BaCO_3$ in aqueous EDTA solution under stirring at the boiling temperature. A citrate solution was also prepared by adding Y_2O_3 and CuO into a solution of citric acid in boiling water further acidified with HNO_3. Both the solutions (pH > 5) were boiled, and cooled to room temperature with their pH adjusted to ≥ 7. The EDTA solution was added to the citrate solution with NH_4OH as diluent when necessary. When the excess NH_4OH was evaporated, a glassy, blue gel was obtained; on vacuum drying, the gel transformed to a blue powder. Calcination of the powder at 850°C/12 h followed by annealing at 400°C led to the formation of YBC with a trace of $BaCuO_2$. A similar investigation on citrate gel preparation in the YBC system reported recently used copper carbonate instead of oxide to obtain the gel [508].

Another citrate method takes advantage of the combustion phenomena [509] in citrates. Nitrate solutions of YBC stoichiometry were diluted and mixed with excess citric acid, with the pH of the solution adjusted to 6.8 by ammonia. A dark blue gel was obtained on vacuum evaporation of the solution. The combustion reaction in the dried gel yielded a dark brown powder which, when calcined in vacuum at ≥ 650°C and then under a flow of oxygen at 900°C, yielded YBC after slow cooling in oxygen.

In a similar work on auto-ignited combustion process [510], nitrate-citrate solutions in the cation ratio Bi : Pb : Sr : Ca : Cu = 1.75 : 0.25 : 2 : 2 : 3 ("2223") and with 1 g.mol of citric acid per g.mol of cations were prepared. EDTA was added dropwise in it until the pH value was 6. The solution was heated to 80°–90°C when it became very viscous and finally turned into a blue-green gel. With continued heating, the gel foamed and caught fire, leading to the formation of an ash. At 800°C/4 h, the product was primarily the "2212" phase, accompanied by "2223" and "2201" phases.

An acetate-citrate method was described by Thomson et al. [511], in which a precursor was prepared by dissolving acetates of Y, Ba and Cu in citric acid and ethylene glycol and heating it to 170°C. The product was diluted by water, but again heated to 80°C to remove the water and obtain a viscous solution. This was dried at ~ 75°C /48 h and ground to a fine powder.

In a recent work [512], a polyester was prepared by adding a 30 wt% malonic acid into an emerald green solution of Ba-acetate dissolved in glycerol, in which nitrates of Y and Cu were added, and heating the precursor thus obtained. By slow heating, excess glycerol was removed and the glycerol-malonic acid polyester decomposed.

Spray pyrolysis has been used by several workers for obtaining cuprate powders. Pebler and Charles [513] obtained close to 100% yield of YBC of 0.1 –1 μm size at 1000°C by the decomposition of a mixed nitrate solution with O_2 as the carrier gas. Kim et al. [514] also used a mixed nitrate solution and decomposed it at 800°–1100°C. A range of 900°–1000°C was confirmed for obtaining single-phase YBC; $BaCO_3$ was obtained below 900°C, while non-superconducting compounds crystallized above 1000°C in air atmosphere. Decomposition of droplets of a stoichiometric nitrate solution (using O_2 as the carrier gas) to essentially YBC at 1000°C was also confirmed by Kodas et al. [515].

Spray pyrolysis has been used by Tohge et al. [516, 517] for the synthesis of Bi-based superconductor powders. In one case, the starting material was a mixed nitrate solution with Bi : Ca : Sr : Cu = 2 : 2 : 2 : 3. When Pb-nitrate was added, the cation ratio of Bi : Pb was 1.8 : 0.2 for Bi-Pb substitution, the rest remaining the same. Unfortunately, the Pb-containing phase was never obtained in pure form. In the Pb-free system, a superconductor (T_c = 80 K) was obtained as the single phase at 800°C under N_2 flow and at 840°C under N_2-O_2. The mean particle size varied from 1.4 μm to 0.4 μm as the solution concentration decreased [516].

In another similar work [517], Tohge et al. used a nitrate solution containing Bi : Ca : Sr : Cu in the atomic ratio 2 : 1 : 2 : 2 for spraying. Nitrogen was used as the carrier gas. The products obtained from a 0.45 mol/lit solution, when calcined precisely at 800°C, yielded phase-pure $Bi_2Ca_1Sr_2Cu_2O_x$ particles. The mean particle size (~ 2 μm) decreased with a decrease in the solution concentration. A carrier gas flow rate of 0.1–1 lit/min was found to be important for obtaining single-phase material.

Alkoxides are less favoured starting materials for preparing superconducting cuprate powders because they are difficult to obtain, and hence, expensive. Unfavourably low or high chemical reactivity of some of the members is also an important reason. Some work has still been done with alkoxides. One example is cited below [518].

Copper methoxide and yttrium isopropoxide were dissolved in 2-dimethylaminoethanol and ethanol to prepare modified alkoxides. Barium metal was added into it to obtain a stoichiometric YBC precursor, which was refluxed in N_2 for 3 h. Hydrolysis was allowed by the addition of water, followed by further refluxing in N_2 for 15 h. The solvent was evaporated at 40°C under reduced pressure. The precursor, calcined at ~ 900°C/3 h, produced single-phase YBC of 1–2 μm size.

Hydroxyapatite

Hydroxyapatite is a member of a group of calcium phosphate and phosphate-hydrates belonging to the family of apatites. Table 4.3 provides a list of the calcium phosphates which relate to the biological applications of apatite [519].

Table 4.3 Apatites of biological importance

Chemical/Mineral Name	Chemical Formula	Ca/P Ratio	Acronym
Brushite	$CaHPO_4.2H_2O$	1	DCPD
Octocalcium phosphate	$Ca_8H_2(PO_4)_6.5H_2O$	1.33	OCP
β-Tricalcium phosphate	$Ca_3(PO_4)_2$	1.50	β-TCP/ Whitlockite
Hydroxyapatite	$Ca_{10}(PO_4)_6(OH)_2$	1.67	HAp
Tetracalcium phosphate	$Ca_4(PO_4)_2O$	2.00	TCPM

Of these, hydroxyapatite is the major inorganic component of the hard tissues, e.g. bones and teeth of various vertebrate animals and human beings. This makes it the most studied among the biologically important phosphates.

Like various multicomponent ceramic powders, hydroxyapatite can also be prepared by "dry" mixing techniques. Calcium phosphates and carbonates are generally mixed stoichiometrically under water vapour to produce HAp. However, these techniques will not be elaborated here.

The simplest method [519–521] of preparing HAp in a relatively controlled way is to mix water soluble sources of calcium, e.g. $Ca(NO_3)_2$ and phosphorus, e.g. $(NH_4)_2HPO_4$ or $(NH_4)H_2PO_4$ in right proportions in an aqueous solution and increase the pH of the system by dropwise addition of NH_4OH at an elevated temperature (80°–90°C) so as to obtain a precipitation. Following the usual practice for other ceramic powders, the precipitate is filtered and washed, and freeze-dried to obtain irregular particles, or spray-dried or spray-pyrolyzed to obtain spherical agglomerates of several micron diameter (Fig. 4.10). Bako and Kotsis [522] have made a detailed study on the conditions of precipitation.

Detailed results on the synthesis of nominally hydroxyapatite powders by spray pyrolysis of solutions obtained from various starting compounds have been recently published by Aizawa et al. [523]. Calcium was provided by nitrate, chloride and acetate salts, each of which was mixed in an aqueous solution (pure water or water acidified with HNO_3/HCl) with $(NH_4)_2HPO_4$, H_3PO_4 and $PO(OCH_3)_3$; the proportions of the chemical pairs were always such as to maintain a Ca/P molar ratio of 1.67 (stoichiometric HAp). The pH of the system generally varied from about 1 to 3.

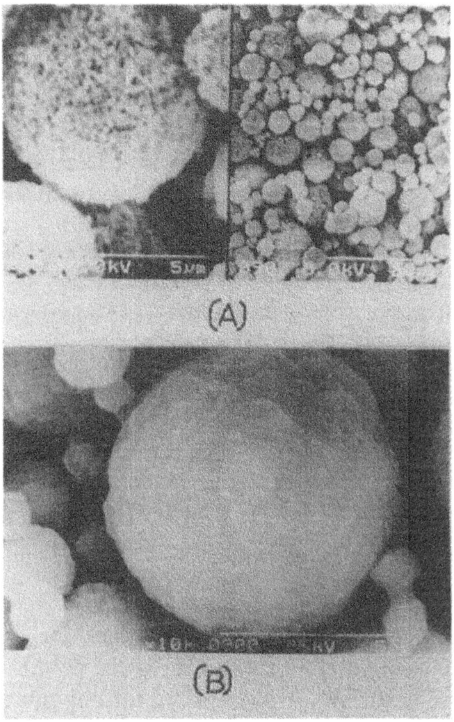

Figure 4.10 Spray-dried hydroxyapatite powder [519]. (A) As-spray-dried; (B) Sintered at 1200°C/2 h. (Reproduced with permission from the Indian Ceramic Society, Calcutta, India).

Spray pyrolysis at 600°C of the nine solutions thus obtained yielded a variety of apatites. Among the precursors, only $Ca(NO_3)_2$–$(NH_4)_2HPO_4$ and $Ca(NO_3)_2$–H_3PO_4 combinations yielded stoichiometric hydroxyapatite; the particle size could go up to 5 μm. The best sintered products were obtained from the $Ca(CH_3COO)_2$–$(NH_4)_2HPO_4$ solution.

Inoue and Ono [524] used $Ca(NO_3)_2$ and H_3PO_4 dissolved in a methanol (77–91 vol%)-water (6.2–20 vol%)-nitric acid solution for spray pyrolysis. The pyrolysis temperature was generally 1100°C, but varied between 800° and 1400°C. The process included the preparation of doped (F, Cl, Na, K, Mg) particles. The product consisted of spherical particles up to ~10 μm in diameter.

Freeze drying of a solution of $Ca(CH_3COO)_2.H_2O$ and $(C_2H_5O)_3PO$ with an excess of P in the system has been described by Hattori et al. [525]. Rapid freezing in liquid nitrogen followed by calcination at 400°–1000°C for 2–6 h showed that the process could yield HAp, but variable quantities of CaO and $CaCO_3$ generally co-existed.

A water-in-oil type emulsion technique has been reported by Deptula et al. [526] for obtaining large (up to nearly 100 μm) spherical particles. $Ca(CH_3COO)_2$ and H_3PO_4 were dissolved in water in the stoichiometric ratio for HAp (water phase). The oil phase was constituted of 2-ethylhexanol and a surfactant. Phase-pure HAp was obtained after drying at 500°C/24 h and calcination at 900°C/24 h.

Ravaglioli et al. [49] describe a method of synthesis using $Ca(OH)_2$ as the source of calcium. This insoluble compound was mixed in water in a pyrex reactor and stirred so as to obtain a dispersion of fine particles. The temperature was raised to ~ 100°C, but again brought down to 20°C when dripping of H_3PO_4 as the source of phosphorus was started. The addition continued for 2 h while the temperature of the system went up to ~ 65°C. The reaction

$$5Ca(OH)_2 + 3H_3PO_4 \rightarrow Ca_5(PO_4)_3(OH) + 9H_2O \qquad \text{... (4.39)}$$

led to the formation of hydroxyapatite, but it required 5 days' digestion and another 2 days' ageing at 37°C in contact with air. Apparently the powders were of very pure quality.

Similar preparative routes have been reported by others [527–529]. Of these, Osaka et al. [527] described the technique in great detail. According to these authors, the ageing (or "ripening") was required for obtaining better crystallinity of the synthesized HAp, and avoiding non-stoichiometry (formation of β-TCP) in the product; the aged samples thus had better thermal stability. While Ravaglioli et al.[49] allowed a slow rise of the reaction temperature from 20°C to ~ 65°C, Osaka et al. [527] found that precipitates prepared at or below 60°C would lead to the formation of β-TCP. They also showed that only a Ca/P molar ratio of 1.68–1.70 could yield a thermally stable (~ 1200°C), single-phase HAp. High Ca/P ratios yielded CaO as an accompanying phase; in case of low ratios, it was TCP. The crystallite size was ~38 nm at 100°C, but was about 15 nm at room temperature.

Calcium hydroxide was also used by Yamada et al. [530] along with $CaCO_3$ (in 1 : 2 wt ratio) as the source of calcium in an unusual synthetic technique. Phytin was extracted from oilless rice bran with HCl solution at 60°C/1 h. The cationic impurities were removed by passing the product through cation exchange resin beds. The calcium compounds were added to the extract of phytin, and the mixture kept at 60°C/1 h under stirring to obtain a precipitate of Ca-phytate. This precursor was finally calcined at 1000°C/3 h to obtain a HAp powder.

The question of formation of carbonate-containing hydroxyapatite (and other apatites) was examined in some detail by Aizawa et al. in a different work [531]. The reactants were $Ca(NO_3)_2$ and $(NH_4)_2HPO_4$, mixed in stoichiometric quantity in a solution. An additional component was urea (up to 7.00 M concentration in the solution). The solutions were pyrolyzed at 600°C. In absence of urea, the products were β-TCP and HAp. With urea of 1.0–3.0 M concentration in solution, only hydroxyapatite was apparently obtained. However, from zero to 7.00 M urea, the Ca/P molar ratio changed from ~1.66 to 1.73, indicating incorporation of $(CO_3)^{2-}$ groups in the structure of hydroxyapatite, which was confirmed by structural studies. Substitution of both $(PO_4)^{3-}$ and $(OH)^{-}$ by $(CO_3)^{2-}$ was suggested respectively for low and high concentrations of urea. The specific surface area varied accordingly from about 15 to 42 m^2/g.

A mechanochemical method of synthesis of HAp (along with β-TCP) has been recently described by Toriyama et al. [532], in which the starting materials were brushite and anhydrous $CaCO_3$ in a slurry with a powder content of 13–20%. The milling time varied from 3 to 48 h depending on the type of mill and the type of container (teflon, ZrO_2). Under limited conditions, poorly crystallized HAp was obtained which, on calcination, turned to well-crystallized HAp with β-TCP. On calcination, the particle diameter increased from < 12 nm to 43 nm, while the specific surface area decreased from 169 to 44 m^2/g.

Zircon

Zircon, $ZrSiO_4$ crystallizes in the tetragonal system. It is a natural mineral generally obtained as a constituent of beach sands along the coast lines of some countries. The dissociation of zircon and the release of zirconia (Section 4.1.2) for ceramic applications is a continuing industrial activity. On the other hand, formation of zircon from ZrO_2 and SiO_2 or other suitable starting materials is also an old [31] and important process for the pigment industry. Various synthetic processes are known for the synthesis of zircon for pigment and other applications. Some representative examples are cited below.

Monoclinic zirconia and low quartz powders have been mixed with V_2O_5 and heated up to 1100°C for obtaining V-zircon pigments [533]. In this process, the oxides of vanadium and zirconium first reacted at ~730°C to generate a vanadate, which then reacted with silica to give rise to vanadium zircon with crystallographic evidence of vanadium entering into the zircon structure.

$$ZrO_2 + nV_2O_5 \rightarrow ZrO_2.nV_2O_x \; (n = 4 \text{ or } 5) \quad \ldots (4.40)$$

$$ZrO_2.nV_2O_x \rightarrow \text{V-Zircon} + V_2O_5 \quad \ldots (4.41)$$

A detailed investigation was reported by Demiray et al. [534] where various sources of ZrO_2 (ZrO_2 powder, $ZrOCl_2$) and silica (silicic acid, $SiCl_4$), vanadium oxides of various valencies, and some additional components (e.g. fluorides as flux) were used for the synthesis of vanadium zircon blue pigments. The precursors thus prepared were heated in air or a controlled atmosphere in the temperature range 1000°–1330°C for 24–72 h. In most cases, zircon was found to crystallize.

Considering the difficulties of obtaining pure zircon via solid-solid reactions, most investigators resorted to solution routes for the synthesis. One such investigation was reported by Kadogawa and Yamate [535], who used several first transition metal ions (Co, Mn, Cu, Ni, Fe) in the form of chloride and nitrate salts in sols prepared out of TEOS (tetraethyl orthosilicate) and zirconyl chloride. They demonstrated that heat-treatment at 1350°C/0.5 h of gels out of the two-component sols, when not doped with one of the above ions, produced no zircon. In presence of Ni^{2+}, only zircon was obtained at 1200°C/0.5 h.

The continued interest in the synthesis of new zircon-based pigments is evident from the recent investigations of Monros et al. [32, 536]. Using both sol-gel and conventional ceramic routes, these authors developed various solid solutions involving the incorporation of the vanadium ion in the $ZrSiO_4$ host structure :

(i) A soft blue stain (sol-gel) involving V^{4+}.

(ii) A green stain (sol-gel, ceramic) involving both V^{4+} and V^{5+}.

(iii) A turquois blue stain (sol-gel, ceramic) involving V^{4+} and F^- in $ZrSiO_4$.

Oheim et al. [33] have recently reported a sol-gel method in which a praseodymium solution was first prepared by dissolving Pr_6O_{11} in HNO_3 and re-dissolving the crystalline product obtained therefrom. A stabilized solution was prepared by adding acetylacetone and diluting with methanol or isopropanol. This Pr-solution was added to an acetic acid solution of zirconium propoxide, followed by the addition of TEOS. A gel was obtained within an hour, which was dried at 200°C and calcined at 1100°–1500°C with or without $ZrSiO_4$ seeding or NaF/NaCl mineralizers. The best results were obtained with mineralizers at 1250°C, when zircon was the predominant phase.

Vilmin et al. [537] presented an elaborate scheme of synthesis of zircon through the crystallization of gels. Two types of gel were used : (a) structurally diphasic gels, which were made out of sols of TEOS and $ZrOCl_2$ seeded with hydrothermally synthesized crystals of $ZrSiO_4$, and (b) compositionally diphasic gels prepared from commercial silica sol and a zirconia sol containing monoclinic or cubic zirconia particles. It was shown that the seeded gels crystallized to zircon (not phase-pure) at least 200°C below the unseeded ones.

Kanno [538] prepared zircon sols from $ZrOCl_2.8H_2O$ dissolved in hot ethanol and TEOS mixed with it under stirring at room temperature. The gels obtained from these sols were dried and milled in a teflon pot with teflon media for 48 h; a part of the gel was not milled. The unmilled samples yielded no zircon up to 1650°C, but a substantial proportion of it at 1700°C. On the other hand, a significant amount of zircon formed from milled gels already at 1580°C.

Mori et al. [539] indicated that the formation of Zr-O-Si bonds in the mixed precursors and a control on the concentration of Zr^{4+} facilitated the formation of zircon. Starting from compositions prepared out of $ZrOCl_2$ and colloidal SiO_2, these authors obtained single-phase zircon at 1300°C.

The route followed by Haaker and Ewing [540] in preparing a zircon sol had the following steps :

(i) Dissolution of a zirconyl nitrate in methanol/H_2O solution.

(ii) Dissolution of TEOS in methanol.

(iii) Addition of the Zr-solution into the Si-solution.

(iv) Dropwise addition of the mixed Zr-Si solution into a 2 : 1 mixture of H_2O and concentrated ammonia solution.

The gel obtained from this hydrolyzed product was dried at 90°C and calcined at 700°C/0.25 h. Heating of the product of calcination at ~1475°C/24 h caused the crystallization of zircon with rather coarse (commonly ~ 5 μm) particles.

In some investigations, only alkoxides (e.g. zirconium isopropoxide and TEOS) have been selected as the starting materials. In one such case, Kobayashi et al. [541] optimized the conditions of sol preparation and seeded them with 1.5 wt% $ZrSiO_4$ crystals for the conversion of the gel products at high temperature. It was found that at 1200°C/2 h, the yield was low; it increased at 1250°–1350°C, and at 1500°C/2 h, an almost single-phase zircon powder (0.3–0.6 μm) was obtained.

In a recent work, Shi et al. [56] prepared highly pure (~ 99.5%) zircon powder from zirconyl chloride and fumed silica. The particulate sol produced from the above was converted to a gel by the addition of NH_4OH. When the gel was seeded with 7 mol% natural zircon, almost phase-pure zircon was obtained by a two-step heating : 700°C/2 h + 1400°C/2 h; the slight impurity was caused by t-ZrO_2. When the temperature of the second step was increased to 1500°C, a single-phase

$ZrSiO_4$ powder was obtained. The temperature of crystallizing 100% zircon has been recorded to be less, i.e. 1300°C when ethanolic solutions of $ZrOCl_2$ and TEOS were mixed and added dropwise to a NH_4OH solution to obtain an immediate precipitate [542].

Zircon has also been crystallized hydrothermally [543] from $ZrOCl_2$, TEOS or TMOS (tetramethyl orthosilicate), water and/or alcohol in teflon-lined autoclaves at 150°C/6 h or 200°C/4 h.

Cordierite

Cordierite, $2MgO.2Al_2O_3.5SiO_2$ or $Mg_2Al_4Si_5O_{18}$ is often described in the literature in terms of two polymorphic forms : the μ- or low temperature (orthorhombic) phase and the α- or high temperature (hexagonal) phase [544] with a transformation temperature of around 900°C. It has been found later that the two phases were indeed linked structurally through an array of intermediate stages depending on the ordering of Al and Si in the framework [545].

Much of the modern work on the powder preparation of cordierite uses different varieties of the sol-gel and related techniques. Some examples are discussed below. Kazakos et al. [546] prepared three kinds of sol, with or without α- cordierite seeding. They were :

(i) Monophasic sol : Al-nitrate and Mg-nitrate solutions in alcohol were mixed together, and tetraethyl orthosilicate (TEOS) added to it under stirring. The sol-gel conversion took place at 60°C.

(ii) Diphasic sol : Boehmite powder was dispersed in water and peptized with HNO_3; a commercial silica sol was added to it. A Mg-nitrate solution was mixed with it under stirring to obtain a gel.

(iii) Triphasic sol : Sols of the three cations were mixed together to obtain a dispersion.

All the gels were dried at 100°C and calcined at 400°C and higher temperatures. The seeded samples yielded single-phase cordierite at relatively low temperatures (e.g. 1100°C for the monophasic gel, as compared to 1200°C required in case of unseeded gels).

Extensive work on various aspects of sol-gel synthesis of cordierite powder has been reported by Okuyama and associates. Brief descriptions of the methods will be provided below. The main starting materials used by Okuyama et al. [547–549] were TEOS, Al-s-butoxide and Mg metal. The TEOS was dissolved in ethanol and partially pre-hydrolyzed (H_2O : TEOS molar ratio required to be ~1.0 and above to obtain the desired final product). It was then mixed with an ethanolic solution of the Al-alkoxide and refluxed. Metallic Mg was added to it under further refluxing to obtain a mixed sol.

The above sols were diluted in ethanol and hydrolyzed with an aqueous solution of NH_4OH. After long ageing times (one day to several days), amorphous products were obtained which were calcined to 1050°C for obtaining cordierite.

The end product of the hydrolysis-condensation reactions was found to be highly dependent on the alkoxy-group of the precursor and the nature of the solvent. The alkoxy-groups in the mother precursor described above were exchanged for other groups by dissolving it in different alcohols and refluxing. A clear region of powder formation was demonstrated (Fig. 4.11)

Exchanged alkoxy groups	Solvent			
	CH_3OH	C_2H_5OH	2-C_3H_7OH	2-C_4H_9OH
CH_3	G	G	G	G
C_2H_5	P	P	G	G
C_3H_7	P	P	P	G
C_4H_9	P	P	P	G

Figure 4.11 Effect of alkoxy-groups and solvents on powder generation from a "cordierite" sol [548] (G = gel, P = powder).

Out of this, again, the butoxy-precursor, dissolved in methanol, gave rise to discrete, spherical particles (0.2 μm) of cordierite. The experimental results have been described in detail by Okuyama et al. [548]. A rapid addition of the water of hydrolysis at a high pH (11.7 or more) was found suitable for proper stoichiometry and purity [549].

All-alkoxide starting materials have also been used for obtaining cordierite powder by Suzuki et al. [550]. Three different starting materials were prepared and their performance compared :

(i) TEOS, Al-isopropoxide and Mg-ethoxide were dissolved in methanol and refluxed at 90°C, after which water (H_2O/TEOS = 20) was added at room temperature. The product was dried at 100°C .

(ii) TEOS dissolved in ethanol was partially hydrolyzed by refluxing at 70°C/50 h in presence of water and catalytic quantity of HCl. A separate solution of Mg-ethoxide and Al-isopropoxide in butanol was refluxed at 95°C. The solutions were mixed and water (H_2O/TEOS = 20) added to the clear solution.

(iii) Another precursor was obtained in exactly the same way as in (i) above, except that the partial hydrolysis of TEOS continued for 100 h.

All the three powders remained amorphous up to 900°C. Powders from precursors (i) and (ii) yielded an assemblage of β-quartz solid solution, spinel and unknown phases up to 1400°C; however, the precursor (iii) generated phase-pure cordierite from 1000°C onwards.

Gensse and Chowdhry [551] used a low-water (H_2O/TEOS = 2) Si-alkoxide solution for partial hydrolysis and mixed it with solutions of Mg- and Al-s-butoxide in amyl alcohol (1 wt % acetic acid added in case of the former). The mixture was slowly added to a solution of ammonia-saturated water at 70°C for 24 h under stirring. For better reaction (leading to 100% yield), ammonia was also bubbled through the mixed solution. The powder thus produced crystallized to μ-cordierite at 1150°C and α-cordierite above 1350°C.

An elaborate organic gel method has been exploited by Douy [366] (see the discussion on mullite, Section 4.1.8) for obtaining cordierite powder. Two routes were followed : organic gel preparation with or without citric acid. Powders of cordierite composition were obtained through both the routes; on calcination at ~ 1000°C, they gave a mixture of μ-(low) and α-(high) cordierite or only α-cordierite under suitable conditions.

Cordierite has been also synthesized [552] by the combustion technique from two precursor mixtures :
(a) aqueous mixture of Mg-nitrate, Al-nitrate, fumed silica, NH_4-nitrate and urea.
(b) aqueous mixture of MgO particles, Al-nitrate, fumed silica, NH_4-nitrate and urea.

Both the mixtures were introduced into a furnace maintained at 525°C; ignition took place in a short while and the synthesis of cordierite was complete in 5 min. The products were low-density (~ 0.15–0.25 g/cm^3) masses containing 10–15 μm powders and high (~ 40–140 m^2/g) specific surface area.

Oxides for Nuclear Applications

The conventional method of preparing fine oxide powders for nuclear fuel fabrication consists of mixing (for multicomponent oxides) and grinding. The major drawback of the process lies in the handling of the fine powders which create radiotoxic dust hazards. Further, due to the lack of a free-flowing nature of the fine powders, remote and automated fabrication becomes difficult [553].

Sol-gel processes [157, 554, 555] have been successfully applied for the preparation of free-flowing, spherical nuclear oxide powders (single and multicomponent, e.g. UO_2, PuO_2, ThO_2-UO_2, UO_2-PuO_2) in the size range of about 50–2000 μm [555] with a high degree of chemical homogeneity. The gel spheres are converted to oxide spheres at considerably low temperatures with controllable porosity and hence, relative density. Such spherical oxide powders of optimally large size eliminate dust hazards during handling.

An early method involved the formation of concentrated aqueous sol droplets, dehydration of the sol droplets to gel microspheres and finally, conversion of the gel microspheres to oxide microspheres [554, 555]. Compared to the above, the "ammonia gelation" processes (either external or internal) are now more popular for the preparation of such powders [553, 556]. The external gelation process involves ammonia gas or ammonium hydroxide as the gelling agent, while the ammonia generated via the decomposition of an amine, e.g. urea or hexamethylenetetramine (HMTA) causes the internal gelation of the sol droplets. Using these general methods, microspheres of UO_2, PuO_2 and their combinations, as also the corresponding non-oxides have been prepared [553]. A typical reaction involved in the formation of uranium oxide gel microspheres is [553] :

$$UO_2(NO_3)_2 + 2NH_4OH \rightarrow UO_3.xH_2O.yNH_3 + 2NH_4NO_3 \quad \text{... (4.42)}$$

The gel microspheres are calcined at 500°–600°C in air, as also under controlled atmosphere when necessary. Figure 4.12 shows spherical UO_2 prepared by the sol-gel method.

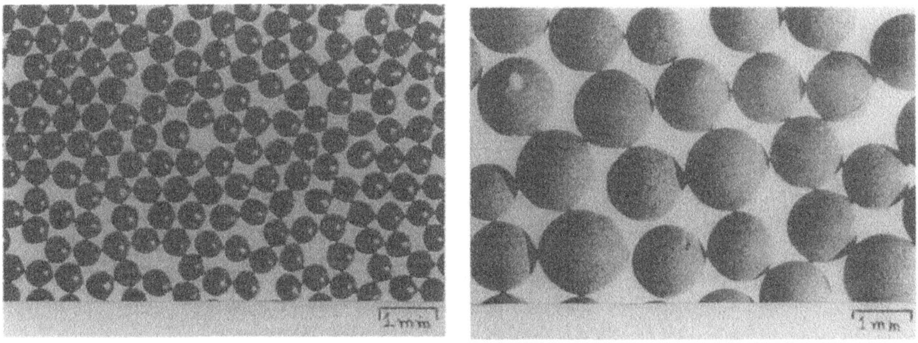

(A) (B)

Figure 4.12 Spherical UO_2 powder obtained by the sol-gel method : (A) by external gelation; (B) by internal gelation (kind courtesy : Dr. C. Ganguly, Central Glass and Ceramic Research Institute, Calcutta, India).

4.2 NON-OXIDE POWDERS

4.2.1 Silicon Nitride

Silicon nitride (Si_3N_4) is generally obtained in two modifications, the α- and β-types, both crystallizing in the hexagonal system. They are considered to be the low and high temperature forms respectively.

Four techniques of preparation of Si_3N_4 powders are known to have been commercially exploited [80, 188]. They are described below in brief.

(a) Nitridation of elemental silicon or "direct nitridation". In this method, silicon powder is heated in a nitrogen or ammonia atmosphere in presence of iron as catalyst at 1200°–1400°C :

$$3Si + 2N_2 \rightarrow Si_3N_4 \qquad \ldots (4.43)$$

This is a low-cost technique, but impurities (Fe, Al, Ca, Mg etc.) from the silicon powder and the post-synthesis pulverization may make the powder contaminated, requiring acid leaching. Figure 4.13 shows the morphology and size of a commercial powder.

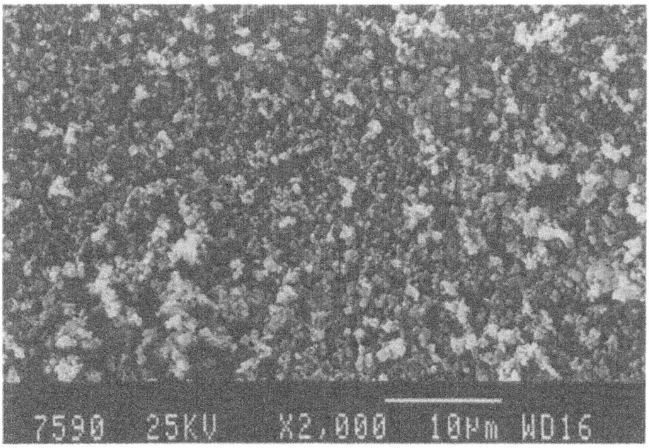

Figure 4.13 Scanning electron micrograph of a commercial (grade M11) silicon nitride powder (kind courtesy : H. C. Starck GmbH & Co., Goslar, Germany).

(b) Vapour-phase ammonolysis of silicon tetrachloride. Vapour phase ammonolysis of $SiCl_4$ (diluted with N_2 or NH_3) at 1100°–1400°C can produce pure and sinterable Si_3N_4 powder with high (> 95%) α-phase content :

$$3SiCl_4 + 4NH_3 \rightarrow Si_3N_4 + 12HCl \qquad \ldots (4.44)$$

$$3SiCl_4 + 16NH_3 \rightarrow Si_3N_4 + 12NH_4Cl \qquad \ldots (4.45)$$

The necessary heat can be supplied by a DC plasma (CVD), RF plasma or other sources. The method generally produces high-purity, ultrafine (30–40 nm) powders, with high residual chlorine but low Si (< 100 ppm). The reaction product is at least partially amorphous and requires calcination at elevated temperatures, which also increases the particle size to 200–300 nm. The powder morphology shows drastic variations from loose agglomerates (amorphous) to fibre- or rod-like particles (partially crystalline).

(c) Thermal decomposition of silicon diimide. When silicon tetrachloride and ammonia react at ambient or lower temperatures, silicon diimide is obtained under controlled conditions:

$$SiCl_4 + 6NH_3 \rightarrow Si(NH)_2 + 4NH_4Cl \qquad \ldots (4.46)$$

$SiCl_4$ is allowed to react at the interface of liquid ammonia and an organic solvent so as to control the production system [188, 557]. The diimide is washed with liquid NH_3 for separating NH_4Cl and finally calcined at 1000°C to obtain amorphous Si_3N_4 particles.

$$3Si(NH)_2 \rightarrow Si_3N_4 + 2NH_3 \qquad \ldots (4.47)$$

The imide-derived Si_3N_4 powders, available in several commercial grades, are 0.2 ~ 0.7 μm in average size and contain about 1 wt% oxygen, but very little cationic impurites.

(d) Carbothermal reduction of silica. In this process, silica powder and excess carbon are allowed to react in the temperature range of 1300°—1700°C in nitrogen; silicon monoxide is formed, which reacts in the vapour phase with nitrogen to yield silicon nitride.

$$3SiO_2 + 3C \rightarrow 3SiO + 3CO \qquad \ldots (4.48)$$

$$3SiO + 3C + 2N_2 \rightarrow Si_3N_4 + 3CO \qquad \ldots (4.49)$$

Any unreacted carbon has to be removed by oxidation to CO_2 after moderate heating at 500°C in air. The process is economically attractive in the sense that it can use natural sources of silica when very high purity is not required. The commercial powders are about 1 μm in average size and contain about 2 wt% oxygen [188].

Many variants of the above commercial processes have been reported in the literature. Some examples are discussed below.

Nitridation of Si powder was carried out in an alumina tube reactor by Qitao et al. [558]. Gas mixtures of N_2, H_2 and Ar in different proportions were employed to allow nitridation, remove oxygen from the surface layers and control the reaction rate. Small quantities (0.5–0.9 wt%) of Fe were also used for catalyzing the removal of silica layers from the silicon particle surfaces. The reaction took place at 1350°–1450°C for 45–63 h. The particles were obtained in the size range of 1–6.5 μm with an α-content of ~ 82–89%. The α-content decreased with increase in the temperature of nitridation, as also an increase in the Fe-content (> 0.5 wt%), but increased when the silicon powder was thoroughly ground (in a WC vibration mill for 24 h).

A combustion synthesis process, described by Hirao et al. [559], used at least 99.9% pure Si (< 5 μm) and > 99.9% pure Si_3N_4 (~ 0.1 μm) particles for reaction with nitrogen. The proportions were given by

$$3(1-x)\ Si + 2(1-x)\ N_2 + xSi_3N_4 \rightarrow Si_3N_4 \qquad \ldots (4.50)$$

The addition of Si_3N_4 was necessary mainly to decrease the adiabatic combustion temperature. The reaction took place at ~ 4100°–2300°C under a nitrogen pressure of 10 MPa. A typical reaction product consisting of 80–90% α-Si_3N_4 had an average particle size of 0.52 μm, a specific surface area of 3.67 m^2/g and ~ 2.5 wt% O_2.

A hybrid (RF/DC) thermal plasma reactor was used by Lee et al. [560] for the reaction of $SiCl_4$ and NH_3. A multistage injection system with a water-cooled ring slit was a special feature. Silicon tetrachloride was injected into an Ar/H_2 arc jet, while NH_3 with H_2 was injected into the plasma through the multistage slit. Multi-stage injection led to the formation of amorphous Si_3N_4 (10–30 nm) with low NH_3/ $SiCl_4$ molar ratio.

A relatively high production rate of ultrafine Si_3N_4 powders, using a DC thermal plasma reactor, has been reported recently [561]. In this system, $SiCl_4$ was injected into the reactor near the argon plasma and NH_3, further downstream (tail end). The amorphous spheroidal powder of 30–60 nm size contained 3–5 wt% oxygen and crystallized around 1500°C to mainly α-Si_3N_4.

Silane has been used as another source of silicon for vapour phase reaction with NH_3 for obtaining Si_3N_4 [94, 184, 186, 562, 563]. The best conditions of synthesis [94, 562, 563] were found to be :

(i) NH_3/SiH_4 molar ratio about 11.6;

(ii) total gas flow rate 400 cm^3/min; and

(iii) reaction temperature 650°C.

Sahu et al. [94] found the nitride particles to be stoichiometric with 1.4 wt% O_2, equiaxed and nanosized; however, up to 1100°C, they were only 5–10% crystalline. Similar results have been obtained by Janiga et al. [563].

Low quartz is the most common source of silica for the synthesis of Si_3N_4 by the carbothermal reduction route. Zhang and Cannon [564] employed an unusual source, namely, monosized spherical particles of amorphous silica prepared by the so-called Stöber technique (Section 4.1.5). An advantage was that the surface area and size of the silica particles could be changed by changing the experimental variables in the Stöber technique. A suspension of this silica of different surface areas was made in water and starch added to it. In a variation of the method, the silica was dispersed in isoamyl alcohol and carbon added to it. In the former case, the suspension was treated at 300°C in nitrogen, while in the latter, it was just dried.

Nitridation of the SiO_2–C mixture took place at ~1400°C under flowing nitrogen (10–100 ppm of O_2 and H_2O). After nitridation, the excess carbon was burnt off at 600°C/8 h in air. For the SiO_2-starch combination, nearly 90 wt% yield (mainly α-Si_3N_4) was obtained after 14 h nitridation with a C/SiO_2 ratio of 16. The average size of the particles was 1.2–7 μm.

Apart from the classical routes, Si_3N_4 particles have also been prepared by another route, namely, precursor decomposition. Vapour phase pyrolysis of tris(dimethylamino)silane, $HSi(NMe_2)_3$ has been studied by Shen and Chang [565] who suggested this to be a potential route for obtaining fine particles of Si_3N_4 at a reasonable cost. The pyrolysis was carried out at 700°–1650°C under vacuum or controlled atmosphere in an induction-heated alumina tube reactor, when the vapour of the precursor was introduced with or without the help of a carrier gas (H_2).

The particles always contained free carbon (a minimum of about 10 wt% at the lower reaction temperatures of 700°–1000°C). They were also amorphous. Such products therefore required two post-synthesis heatings : (a) for carbon removal, which was done only partially by heating under low oxygen atmosphere at 600°–800°C, (b) for crystallization, which was carried out at 1400°–1600°C under nitrogen atmosphere. Most of the products were mixtures of α- and β-Si_3N_4 (generally $\alpha > \beta$) and SiC.

Hexamethyldisilazane, $(CH_3)_3$ $SiNH(CH_3)_3$, abbreviated as HMDS, has been found to be a relatively safe liquid precursor for silicon nitride (or carbide) powders. In a recent work [566], HMDS has been atomized into an aerosol and injected into a CW CO_2 laser beam by a carrier gas (Ar + NH_3 or only Ar). The production rate of the powder obtained by the thermal decomposition of HMDS increased with the flow rate of the carrier gas. The as-prepared powder was amorphous with a specific surface area exceeding $100m^2/g$. The C/N ratio of the powder varied from 0.2 (with Ar + NH_3) to 3.18 (only Ar). Heating at 1500°C under N_2 atmosphere produced well-crystallized particles of Si_3N_4 when the C/N ratio was low.

4.2.2 Aluminium Nitride

Aluminium nitride, AlN, is known to crystallize in the hexagonal system, with a wurtzite-type structure, as also as an oxygen-stabilized cubic structure. The powders can be prepared by a variety of techniques (to be described below), but two of them are generally known to be suitable for commercial exploitation : direct nitridation and carbothermal reduction. For both of these, many elegant laboratory-scale variants have also been developed.

(a) Direct nitridation of aluminium. In this method, aluminium powder is reacted with nitrogen gas (sometimes with 5% NH_3) at 1200°–1300°C :

$$Al \xrightarrow{N_2/NH_3} AlN \qquad \text{... (4.51)}$$

While the raw materials are inexpensive, the method has to reckon with several problems [191]. Aluminium in finely divided state always contains a fine layer of Al_2O_3 which prevents the nitridation reaction to proceed. Further, at temperatures around or above 400°C, a layer of AlN is generated on the surface of Al particles, which also hinders further nitridation. There are also some general problems of keeping the metal particles unagglomerated before use, and keeping the size distribution of the synthesized nitride particles to a narrow range after milling. Proprietary techniques for overcoming some of these problems have been discussed in brief by Narula [194]. Other examples of efforts in these areas are given below.

Aluminium nitride is also synthesized commercially by the so-called "Carbonitridation" technique [191] in which Al metal particles are mixed with carbon particles for reducing the alumina layer and the mixture allowed to react with nitrogen at more than 1100°C. The excess carbon is oxidized at 600°–700°C. Finally, the reaction product is heated at around 1400°C in an inert atmosphere (vacuum or nitrogen) to obtain AlN powder with about 1.0–1.5 wt% oxygen and 0.1 wt% or less carbon [567].

A direct nitridation process has been described, in which Al powder mixed with a lithium salt (chloride, fluoride, carbonate etc.) equivalent to about 0.5–2.0 wt% Li is subjected to nitridation [191, 568]; the lithium salt reacts with the surface alumina on Al particles, giving rise to gaseous products which leave the system.

(b) Carbothermal reduction of alumina. In this process, Al_2O_3 powder, generally crystalline, and carbon particles, e.g. petroleum coke, graphite and lamp black, both of submicron size, are physically mixed by milling and allowed to react with nitrogen or ammonia above 1400°C [569].

$$Al_2O_3 + 3C + N_2 \rightarrow 2AlN + 3CO \qquad \text{... (4.52)}$$

A reaction model for the carbothermal reduction of α-Al_2O_3 has been proposed with the following intermediate steps [569] :

$$Al_2O_3 \rightarrow 2Al + 3/2O_2;\ 3C + 3/2O_2 \rightarrow 3CO\ \text{(slow)};\ 2Al + N_2 \rightarrow 2AlN\ \text{(fast)} \qquad \text{... (4.53)}$$

The synthesized powder is fine (1–2 μm), has a narrow size distribution and does not require milling. The oxygen content is 1 wt% or less, and with effective control, the carbon content can go down to 0.02 wt%. A better contact between the oxide particles and the source of carbon, and hence, a faster reaction rate, has been obtained by using suitable polymers instead of carbon [570].

Some examples will be provided below of reported newer investigations based essentially on the above two techniques.

Aluminium nitride has been synthesized recently by the reaction of vaporized Al with NH_3 or N_2 within an externally heated tube at 1400°–1600°C [571]. Ammonia was found to be more reactive than nitrogen, yielding phase-pure AlN under almost all conditions of experimentation. In a different investigation [572], both NH_3 and N_2 were used in different proportions. With increasing NH_3 content, the primary particle size increased : 0.02 μm (40% NH_3); 0.05 μm (80% NH_3) and 0.1 μm (100% NH_3). With only N_2, AlN was not phase-pure.

A rapid process for manufacturing AlN in pilot scale has been described by Weimer et al. [92] in which fine Al metal and nitrogen gas reacted under heated conditions (1600°C) while passing through a transport flow reactor to yield AlN. The product required only mild milling and a post-treatment at 1000°C/1 h. The oxygen content (~ 1 wt%) decreased to 0.45–0.75 wt% when a mixed Al/AlN feed was used (due to low surface area, i.e. 2–3 m^2/g). The possible explosion hazard in the handling of atomized Al powder was indicated by the authors.

A nitridation technique using an RF plasma has been described by Baba et al. [573]. The plasma gas ($Ar\text{-}N_2$ mixture) was introduced from the top in a vertical reaction chamber filled with Ar after evacuation. The reactants, Al powder and NH_3, were introduced at the tail of the generated plasma flame. The powder obtained was single-phase AlN with an average size of 60 nm, 30 m^2/g specific surface area, metal impurity content ~ 100 ppm and oxygen content 2–5 wt%, the latter being a point of concern.

An interesting DC plasma synthesis of AlN has been reported by Lu and Pfender [574] who used three identical DC plasma torches of 3–30 KW power in a cone-like geometry so that the penetration of particulate matter into the thermal plasma (which had a high viscosity) could be facile. Aluminium powder was fed into the plasma with N_2 as carrier and sheath gas; five cm downstream of the tri-junction of the plasma jet, the deposited product was phase-pure hexagonal AlN. The median size was about 2 μm, obviously that of an agglomerated powder.

The possible effect of the source of aluminium oxide in the carbothermal reduction process has been examined in some detail. It has been shown that a variety of

Al_2O_3 powders can be used as the source of Al without significant changes in the process [569]. An elaborate investigation was made by Tsuge et al. [575] on the relative merits of different sources of aluminium as mentioned below, with the average particle size in parentheses : η-, θ- and α-Al_2O_3 (all ≤ 1 μm); γ-Al_2O_3 (0.005 μm or less); $Al(OH)_3$ (~ 0.4, 4 or 15 μm). These powders were mixed with excess amorphous carbon (0.02 μm), in the ratio C/Al_2O_3 = 4. After heat-treatment of the intimate mixtures at 1550°C for 5 h under flowing N_2, phase-pure AlN was obtained with $Al(OH)_3$ (1–1.60 wt% O_2) and γ-Al_2O_3 (1.5–1.8 wt% O_2).

Apart from the reaction temperature, gas flow rate, gas pressure and C/Al_2O_3 ratio, the particle size and the degree of agglomeration have been shown to depend also on the choice of the raw material [576]. Cho and Charles [577] showed that the reaction rate could be increased by using an $Al(OH)_3$–C mixture.

Various other sources of alumina have been utilized for AlN formation by reduction reactions. Thus, alumina gel particles (obtained from the secondary butoxide) and activated carbon were allowed to react at elevated temperatures in flowing N_2–H_2 mixed gas atmosphere [578]; at 1350°C, only AlN crystallized.

Another more elaborate method of precursor preparation in carbothermal reduction [579] used a sol obtained by dispersing pseudoboehmite particles in a dilute solution of HNO_3. Furfuryl alcohol was dissolved in this sol (~ 15 vol%) when it started undergoing polymerization. The reaction was continued at 105°C for 3 days. A black, solid product (alumina/polyfurfuryl alcohol) obtained in this way was ground into a powder and reduced in nitrogen at 1575°C/6 h. After burning off the excess carbon, 0.2–0.3 μm particles of phase-pure AlN were collected.

Apart from the above, various other preparative techniques have been described for AlN. Some of them are discussed below.

An widely studied preparative technique for AlN is the $AlCl_3$–NH_3 vapour phase reaction. It must be remembered here that NH_3 dissociates at about 125°–225°C at 0.1 MPa, i.e.

$$2NH_3 \rightarrow N_2 + 3H_2 \qquad \text{... (4.54)}$$

and the dissociation hinders AlN formation. For the reaction to proceed in the right direction, therefore, it is necessary that intermediate adducts, i.e. $AlCl_3.xNH_3$ ($x = 1 - 6$) form, which can decompose in the following way above 700°C :

$$AlCl_3.NH_3 \leftrightharpoons AlN + 3HCl \qquad \text{... (4.55)}$$

Nickel et al. [580] used an atmospheric pressure CVD system comprising a quartz reactor, a sublimation vessel for $AlCl_3$ and a collector vessel. Aluminium chloride was vaporized at a rather high temperature of 400°C, and the vapour was carried by 99.999% pure N_2 gas into the reactor kept at 1000°C, but not before a contact with 99.999% pure NH_3 at the entrance of the reactor at 400°C when $AlCl_3.xNH_3$ could form. The adduct decomposed inside the reactor, yieding AlN powder. The product in the collector vessel was amorphous (calcined at high temperatures to obtain single phase AlN) or crystalline (reactor wall deposit). The oxygen content was remarkably low at $\leq$ 0.87 wt%. Riedel and Gaudl [581] re-examined the process and found that in spite of the difficulties of synthesis predicted by thermodynamic considerations, a continuous industrial process with a high percentage of theoretical yield could be achieved in the system.

Kimura et al. [582] used the same reaction for obtaining AlN in a system where sublimated $AlCl_3$ and NH_3 gas were fed into the reactor through two different tubes. Aluminium nitride particles started forming above 400°C as a single phase and reached a size of 0.4–10 μm at 720°C. However, the particle size was an inverse function of temperature, and was in the range of 0.1–0.5 μm at 1190°C.

A pure, wurtzite-type AlN [583] was synthesized from $AlCl_3$ via reaction with NH_3 in a graphite reactor at a relatively high temperature of 1600°–1800°C in flowing N_2. The reaction product contained the adduct $AlCl_3.xNH_3$ and NH_4Cl in addition to AlN; the former were removed by sublimation at 950°C in N_2. Calcination of the residue at 1600°C/2 h probably caused a growth of the particles by agglomeration (average size 0.4 μm). The powder contained low impurities, e.g. C 0.078 wt%, O_2 0.85 wt% and Cl(0.2 wt%); the surface area was about 7.5 m^2/g.

Yet another investigation involving very pure $AlCl_3$ has been reported by Park et al. [584] : the vapour was led to a quartz reactor by 99.999% N_2 and allowed to react with 99.999% NH_3 at 700°–1100°C. The obtained AlN particles increased in crystallinity with increasing reaction temperature, and had an average size of 0.2 μm with spherical morphology. Both crystallinity and yield (up to about 80%) increased with increase in $NH_3/AlCl_3$ molar ratio.

Considering that products of carbothermal reduction and direct nitridation are not sinter-active enough, Seibold and Rüssell proposed a polymer breakdown (Section 3.6.2) technique for producing AlN [197]. In their work, polyiminoalane, $\frac{1}{n}[AlH(NR)]_n$ could be converted to a polymer foam, and finally, a gel. The product was calcined up to 1500°C for up to 4 h in NH_3, N or Ar atmosphere. Crystallization was clearly observed at 850°C, and copious at 1100°C. The micron-sized powder thus obtained had 1–1.5% carbon when prepared under NH_3 or N_2 atmosphere.

Another polymeric precursor to obtain AlN was poly(isopropyliminoalane), the preparation of which has been described by Sugahara et al.[585]. The white polymeric solid was pyrolyzed at 1000°C/2 h under argon flow. The black, amorphous product thus obtained was calcined at 1600°C/2 h in argon to obtain a well-crystallized AlN.

Coffman et al. [586] selected dimethylaminoalane, $AlH_2N(CH_3)_2$ as the precursor of AlN; one of the reasons was its structural similarity with AlN. It was therefore directly pyrolyzed at 470°C. The compound was also polymerized by dissolving in the xylene isomers to prepare a very dilute solution, and the solution refluxed under N_2 for 2.5 h at 136°C. A grey powder (one-fifth by weight of the precursor) was obtained which on heating under N_2 flow to 1200°C/2 h produced hexagonal AlN containing oxygen-stabilized cubic AlN as a major impurity phase.

In an account of reactions involving an alkylaluminium [587], streams of $Al(i{-}Bu_3){-}H_2$ and NH_3 vapours were mixed at 200°C, while the reaction took place at 600°–1400°C. The as-synthesized particles, containing at least 4.5% O_2, were amorphous at lower temperatures and weakly crystalline at higher temperatures. The particle size decreased from 34 nm at 600°C to 17 nm at 1400°C. Calcination of the powder in $H_2{-}NH_3$ atmosphere, obtained at 1200°–1400°C, yielded better crystallization of single-phase AlN (0.03 μm). In a similar work, Adjaottor and Griffin [588] allowed $Al(C_2H_5)_3$ vapour (with N_2 as the carrier gas) to react with NH_3 at 700°C (mean size of powder 0.13 μm) to 1100°C (mean size 0.07–0.23 μm). The inlet concentration of triethyl aluminium was found to be the most important parameter of the reaction.

Kubo et al. [589] considered the alkylaluminiums (as above) to be effective starting materials with controllable vaporization behaviour. They therefore extended the range of starting materials : $Al(CH_3)_3$, $Al(C_2H_5)_3$ and $Al(i{-}C_4H_9)_3$. Liquid alkylaluminiums were vaporized at 30°–110°C and the vapour introduced into a mullite reactor tube maintained at 1050°C. Very high purity (99.9995%) N_2 gas was used as the carrier; NH_3 (99.99%) was also introduced into the reactor at the same time. The yield of AlN was reported to be about 80%. The powders were poorly crystallized AlN (single phase), the oxygen content increasing from trimethyl (4.6 wt%) to triisobutyl (9.9 wt%) aluminium. The agglomerated particles were bimodal in size (probably a true representation of powders collected from different parts of the reactor) : around 0.1–0.15 μm, and also around 5 μm. The primary particle sizes and specific surface areas were $Al(CH_3)_3$: 19 nm/95 m^2/g; $Al(C_2H_5)_3$: 13 nm/145 m^2/g; and $Al(i{-}C_4H_9)_3$:11 nm/175 m^2/g.

The reaction sequences involving alkylaluminiums and ammonia have been discussed in some detail by Interrante et al. [590].

A process of preparing high-purity AlN powder has been described by Huseby [591]; the process which uses a rather unique working principle is described below in its major steps :

(i) Synthesis of high-purity $(NH_4)_3AlF_6$ from 99.9995% pure Al, HF and electronic grade NH_3 (at least 99.998% pure, with 2.5 ppm H_2O and < 1 ppm O_2, the impurities removed through further purification).
(ii) Decomposition of the fluoride precursor to AlF_3 at 600°C.
(iii) Reaction of AlF_3 with NH_3 at 800°–1000°C for 15–60 h.

The particles obtained were mostly rod-shaped, but sometimes with a platelet morphology; the maximum surface area was 2.6 m^2/g. The cation purity of the product was 99.99%, with < 0.2 wt% O_2.

4.2.3 Boron Nitride

Boron nitride, BN is generally obtained in one of the two well-known forms : a soft, hexagonal form (h-BN) akin in its structure to the graphitic form of carbon, and a cubic, sphalerite-like form of high hardness (c-MN) obtained under high pressure-temperature conditions. A third form crystallizing in the rhombohedral system is structurally close to h-BN.

The early and now standard methods of synthesis of BN include reaction of B_2O_3, boric acid, or borax with carbon and nitrogen/ammonia or organic C- and N_2-carriers like $CaCN_2$ and urea at temperatures up to about 2000°C [45, 46]. The product generally contains a significant amount of unreacted matter and therefore, is usually leached and washed for increased product purity, and then milled. Another method is to recrystallize BN by heat-treatment with Li_3N at 1100°–1550°C, and washing the product for the removal of lithium nitride.

The direct nitridation of boron in a nitrogen atmosphere is another well-used technique in which the reaction takes place at 1400°–1900°C. For obtaining high reaction rates, salts like ammonium chloride and thiocyanate are used. Inert substances like calcium carbonate and phosphate have also been used as fillers for increasing the areas of contact between the reactants [45]. Figure 4.14 shows the general appearance of a commercial h-BN powder.

Kalyoncu [45] has described the preparation of precursors (boranes and borazines) which, on decomposition at relatively low temperatures, produce BN. Three reactions have been discussed :

(a) Reaction between BCl_3 and NH_3 at –70°C, leading to the formation of a white powder. The expected products could be $B(NH_2)_3$ or $Cl_3B_3.N_3.H_3$. The powdery material, when heated at 1100°C under nitrogen atmosphere, yielded BN.

(b) Reaction between BCl_3 and NH_4Cl at a reflux temperature of 200°C. The solid product, $Cl_3.B_3N_3.H_3$, was calcined at 1100°C in nitrogen atmosphere to obtain BN.

(c) Reaction between $NaBH_4$ and NH_4Cl at 170°–180°C. The solid product yielded BN after heat-treatment at 1100°C.

Considering that the B : N atomic ratio in borazine, $B_3N_3H_6$ is the same as that of BN, Hirano et al. [592] synthesized this precursor for further processing. Pyrolysis of borazine was performed in a cold seal pressure vessel (similar to that used for hydrothermal synthesis) at 250°–700°C and 100 MPa for 5 min–3 h. The yield was about 60%.

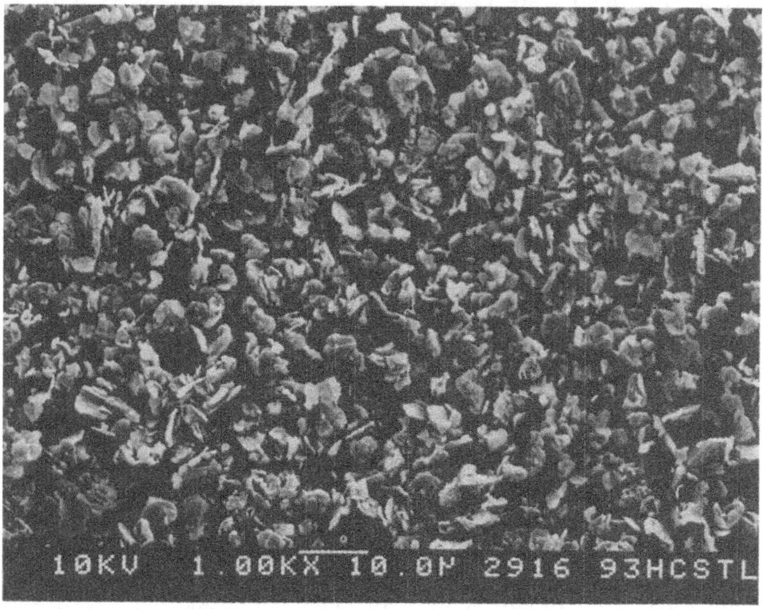

Figure 4.14 Scanning electron micrograph of a commercial (grade A01) boron nitride powder (kind courtesy : H. C. Starck GmbH & Co., Goslar, Germany).

The white solid particles of BN obtained as the residue were practically amorphous, and exhibited properties very sensitive to the preparative conditions. Thus, when produced at 250°C/1 h, the particles were sensitive to moisture and took the form of 4–5 μm spherulites. At 300°–400°C, the product was a mixture of spherulites and fibrils (maximum ≥ 10 μm length, about 1 μm dia); above 400°C, only fibrils were obtained. When pyrolyzed for 3 h, all the products were coalescing spherulites. These products were stable with respect to the environment. The products were converted to cubic BN at 1800°C after adding 20 mol% AlN as catalyst in the charge.

Another investigation of a similar nature was reported by Yogo and Naka [593] who used triammoniadecaborane (TAD), $B_{10}H_{14}.3NH_3$ as the precursor. Under a pressure of 125 MPa in a cold seal type pressure vessel, a solution of TAD in anhydrous hydrazine under N_2, as also pure TAD were pyrolyzed at 400°–650°C for 3 h. The product obtained was a porous mass (composed of nitrogen-deficient, weakly crystallized BN) with large pores, which could be crushed and milled. Hexagonal (and rhombohedral) BN were reported recently to have been produced from "boron-containing solutions" by precipitation in supercritical ammonia at 3–5 GPa and 625°–1325°C [594].

Since its discovery [595], cubic BN has generally been prepared by drastic P–T treatment on hexagonal BN (up to 1700°C and 65×10^8 Pa). Some typical examples are given below.

Hirano et al. [596] used a girdle-type pressure apparatus for converting > 99.8% platelike, highly crystalline BN to the cubic form at 800°–1700°C and 5–7 GPa. Aluminium nitride (20 mol%) was added as a catalyst. An atmosphere of Ar + toluene or N_2 + toluene yielded 100% conversion at 1600°C/6.5 GPa/0.25 h (the transition under Ar started at 1000°C/7 GPa). The transformation did not go through intermediate phases, and there was no diffusion of AlN in BN, ensuring product purity. The particles were ~ 2 μm in size, and sometimes of well-formed tetrahedral morphology. It was observed that under O_2 or air atmosphere, h → c conversion was hindered and AlN converted progressively to Al_2O_3.

Complete conversion of a poorly crystalline (0.1 μm) h-BN to c-BN was obtained [597] at 1200°C but under almost similar pressures as above. Note that the particles became sintered during the conversion. Further, no catalyst was used in this work. Susa et al. [598] used water as a catalyst to obtain the h → c conversion under less drastic conditions (with a cubic anvil and an internal graphite heater). With 40% water in a h-BN paste, the transformation took place at 600°–700°C and 6 Gpa – apparently the lowest known temperature of transformation.

Rapoport and Nadiv [599] mechanically activated h-BN powders by grinding in a centrifugal ball mill to decrease the energy barrier of conversion. This activated powder, along with 30 wt% AlN, yielded up to 70% c-BN when subjected to 1700°C/ 6.6 GPa/0.25 h. The product was treated in molten NaOH + Na_2CO_3 for leaching out h-BN and AlN.

The TAD-hydrazine combination of Yogo and Naka [593], mixed with AlN and heated at 600°C, was also treated by the authors to conditions of 1200°–1400°C/ 6.5 GPa to obtain c-BN.

Two other techniques of obtaining c-BN deserve mention. In one of them, Kobayashi et al. [600] achieved nearly complete yield of c-BN from a mixture of elemental boron and urea by treatment at 1200°C and 6 GPa. Liepins et al. [601], on the other hand, resorted to shock-induced synthesis of c-BN from NH_3BH_3. A localized pressure of about 90 GPa was achieved at > 1000°C by using a detonator-explosive system. The product was essentially agglomerated c-BN, the largest size being around 200 μm.

4.2.4 Silicon Carbide

Silicon carbide, SiC exhibits two modifications : the α-form (various hexagonal and rhombohedral polytypes) and the β-form (cubic). The latter is more frequently reported as the product of vapour phase reactions.

The most important commercial technique of producing SiC powders dates back to late 19th century when Acheson [602] developed a process for the reduction of silica (as silica sand) in presence of coke and sawdust in an electric field.

$$SiO_2 + 3C \rightarrow SiC + 2CO \qquad \text{... (4.56)}$$

The charge was kept between two graphite electrodes in a graphite core; the core temperature was raised to more than 2000°C to obtain the desired product.

The colour of SiC obtained by Acheson process indicates its purity [603] : 99.8% pure material is light green in colour, changing to dark green as the purity falls to about 99.5%; with further fall to 99%, the product shows a black colour. The crude product is separated into batches on the basis of colour (purity), and each batch is crushed and graded for different applications.

Various modifications of the Acheson process, especially in the furnace design, have been carried out [194, 604, 605]. Wei [606] used a mixed charge of silica and carbon (molar ratio 1 : 3 or less) for continuous feeding into a tilted rotary furnace with graphite resistance heating. The better contact between the reactants led to a reduction in the temperature of operation.

Since Acheson's invention and its applications in different modified forms, various combinations of reactants have been used for the synthesis of SiC by carbothermal reduction. A brief but useful summary up to the mid-eighties regarding the choice of new reactants has been made by Wei et al. [607] who extended this by additional work in the following systems :

(i) Colloidal silica and sucrose (CSS process).
(ii) Colloidal silica and pitch (CSP process).
(iii) Methyltrimethoxysilane (MTMS) and phenolic resin (MTMSR process).
(iv) MTMS and sucrose (MTMSS process).

In the processes CSS and CSP, a fumed silica of 14 nm equivalent diameter and a specific surface area of 200 m^2/g was chosen. This was mixed with sucrose or pitch in presence of suitable solvents like toluene. The gel that formed in the process was dried and pyrolyzed. Carbothermal reduction at about 1600°C took place in a static or rotary furnace. A high yield required high C : SiO_2 molar ratios (> 3). The powders were essentially composed of β-SiC, submicron in size (< 0.1 μm) and spherical in morphology, with specific surface area up to 25 m^2/g. The high-carbon compositions required for high yield also left behind a high level of residual carbon (~ 5%). The oxygen content was 3600 ppm for CSS and 4600 ppm for CSP process.

In the processes MTMSR and MTMSS, the source of silica (MTMS) was hydrolyzed with ammonia-water. Sucrose was dissolved in water, while the phenolic resin was dissolved in acetone. The obtained gel was freeze-dried at –50°C and a vacuum of 6.5 KPa. The dried gel was pyrolyzed at 500°C (for sucrose) or 800°C (for the resin) in argon atmosphere. The conversion to SiC powder followed the same route as that for CSS and CSP processes. MTMS-based processes yielded powders with high degrees of excess carbon and silica; removal of the former required heating at 600°C in air for oxidation, while that of the latter required washing with 20% HF. The powders were agglomerates (6 μm) of primarily β-SiC particles. The dark green colour indicated their impure nature (15 wt% C, 2 wt% SiO_2 before purification).

Martin et al. [608] have recently used sugar in a 15 wt% silica sol which was freeze-dried and heated under inert conditions at 1300°–1750°C. The SiC particles were nanosized with O_2 content down to 1% at the highest temperature.

Cerovic et al. [609] utilized the following procedure for obtaining β-SiC powders by the sol-gel route. A silica sol with 3.7 wt% SiO_2, stabilized at pH = 10.3, and saccharose or activated carbon were used as the starting materials; the C/SiO_2 ratio varied from 3 to 5. Boric acid was added as a catalyst. Gelation took place under heating and stirring. Carbothermal reduction of the dried and ground gel particles was carried out under an argon flow up to 1550°C/3 h. With saccharose as the source of carbon (C/SiO_2 = 4), crystallization started relatively early (1300°C); with activated carbon (C/SiO_2 = 3), the onset temperature was 1400°C. In both the cases, 95–96 wt% β-SiC with a specific surface area of around 8-10 m^2/g was obtained.

Precursors like ethyl silicate (partially polymerized TEOS) in combination with phenolic resin have been used for synthesizing about 20–100 nm SiC powders [610]. Toluenesulphonic acid and water have been used to promote the overall reaction. The intermediate steps involved the formation of the suboxide :

$$SiO_2 + C \rightarrow SiO + CO \qquad \text{... (4.57)}$$

$$SiO + 2C \rightarrow SiC + CO \qquad \text{... (4.58)}$$

The volatility of SiO (the formation is known in carbothermal reduction) apparently led to the presence of excess unreacted carbon in the product, but no residual SiO_2 above 1600°C.

In a plasma-assisted reaction reported by Kong and Pfender [611], silicon monoxide reacted with CH_4 at > 10,000°C to generate β-SiC particles of 2–40 nm size. Plasma heating has also been used by other workers [612] to obtain β-SiC from TEOS or $(CH_3)_2Si(OC_2H_5)_2$.

Various vapour-vapour reactions have been studied for the production of SiC powders [67]. Laser and plasma sources have been generally used for supplying the necessary heat. Narula [194] has listed various reactions involving silicon tetrachloride. One example is given below.

Allaire et al. [613] used a DC plasma system to obtain SiC powder by the gas phase reaction between silicon tetrachloride and methane. The tetrachloride (99%) was vaporized at 150°C in a flash boiler and carried by argon into the 10 KW plasma system. Methane was carried by hydrogen further downstream. A vacuum pump evacuated the system before each reaction and also collected the particulate product (β-SiC) in a filter; the product had 2.5–4 wt% free carbon and a C/Si ratio of 1.03–1.05.

Extensive investigations have been carried out on vapour phase reactions involving silane, SiH_4 or its derivatives. Haggerty and associates [e.g. 184, 186] studied some laser-assisted reactions, e.g.

$$2SiH_4 + C_2H_4 \rightarrow 2SiC + 6H_2 \qquad \text{... (4.59)}$$

$$SiH_4 + CH_4 \rightarrow SiC + 4H_2 \qquad \text{... (4.60)}$$

The laser system of Haggerty has been discussed briefly in Section 3.5.2. Lihrmann and Cauchetier [614] pursued this method for studying the reaction between SiH_4 and C_2H_2. The cross-flow of the reactants generated a flame (1300°–1800°C) assisting the reaction. The yield of the fine (15–50 nm) SiC powders was > 99 mol%.

Vogt et al. [20] synthesized SiC in an argon thermal RF plasma. The reactant gas mixtures (SiH_4 : CH_4 generally 1 : 1.5–2.2 standard lit/min) were introduced into the tail flame of the argon plasma. The particles left the system through a water-cooled quenching chamber. They were 20–25 nm in diameter, and had a surface area of around 80–100 m^2/g. The C/Si molar ratio could vary in the range 0.92–1.3. An RF plasma has also been used by Hollabaugh et al. [615] for obtaining SiC from SiH_4 and CH_4. Kondo and Saiki [616] used a multi-DC plasma for the same reactants (including B_2H_6) to obtain B-doped SiC powders.

Chemical vapour deposition techniques have been extensively used by Kato and associates [617–620] for the synthesis of ultrafine β-SiC powders via different starting materials, e.g. $(CH_3)_4Si–H_2$ (1200°–1400°C), $SiH_4–CH_4–H_2$ (1000°–1400°C) etc.

Investigations have also been carried out on the thermal decomposition of silane derivatives for obtaining SiC. Thus, methylsilane (CH_3SiH_3) was thermally decomposed [621] by passing it into an induction heated graphite reactor with Ar as the carrier gas. The decomposition started above 435°C and SiC started forming at 900°–950°C. The yield was proportional to the reaction temperature, and could go to more than 95% at 1800°C. The powder was spheroidal in morphology, with < 0.1–0.5 μm size. Endo et al. [622] decomposed tetramethyldisilane, $H[Si(CH_3)_2]_2H$ at 700° ~ 1400°C under an inert atmosphere to obtain highly pure (metals < 10–24 ppm) β-SiC particles (5 ~ 200 nm, depending on the experimental conditions). No free carbon was detected in the product.

Elemental silicon can be used in all forms (solid, melt, vapour) as a reactant for obtaining SiC at 900°–1570°C [194]. A recent work by Guo et al. [623] reports the preparation of ultrafine SiC powders by the reaction between Si and CH_4. Spherical β-SiC particles of ~10–80 nm size were produced by Ando and Uyeda [624] by a reaction between a carbon rod and a silicon block heated under an argon atmosphere.

4.2.5 Boron Carbide

Commercially, boron carbide (the usual chemical formula is B_4C, though the B/C ratio has been suggested to go as high as 10.5) is produced from a mixture of boric oxide and carbon in the ratio $B_2O_3/C \simeq 2.7$ or higher in an arc furnace at a temperature ranging between 2200° and 2500°C. The strongly exothermic reaction

$$2B_2O_3 + 7C \rightarrow B_4C + 6CO \quad \text{... (4.61)}$$

has been reported to proceed in two stages [625, 626] :

$$B_2O_3 + 3CO \rightarrow 2B + 3CO_2 \quad \text{... (4.62)}$$

$$4B + C \rightarrow B_4C \quad \text{... (4.63)}$$

The above reactions do not show the exact material balance; one of the reasons may be that the product obtained is not stoichiometric ($B_{4.3}C$). The melted carbide thus produced in the inner zone of the batch is crushed, milled and acid washed for powder size grading and purification, and the rest is recycled. Note that this material is usually B-rich. Weimer et al. [627] used a B_2O_3/corn starch combination to obtain 0.1 μm particles at 1900°C.

Magnesiothermal reduction of B_2O_3 [625, 626, 628] involves formation of boron carbide in presence of Mg metal :

$$2B_2O_3 + 6Mg + C \rightarrow B_4C + 6MgO \qquad \text{... (4.64)}$$

This strongly exothermic reaction has been generally carried out at 1000°–1200°C. For the separation of the unreacted Mg or Mg-borides, the product has been washed with H_2SO_4 or HCl, followed by cleaning with hot water. The product was stoichiometric, but required vacuum heating at 1600°C to remove the impurities.

A similar reaction involved aluminium metal particles [628] :

$$2B_2O_3 + 4Al + C \rightarrow B_4C + 2Al_2O_3 \qquad \text{... (4.65)}$$

The starting materials were mixed and cold-pressed into cylinders, and then ignited with a heated tungsten coil under Ar atmosphere. The reaction, which was essentially of the combustion-type, occurred in two steps :

$$2B_2O_3 + 4Al \rightarrow 4B + 2Al_2O_3 \qquad \text{... (4.66)}$$

$$4B + C \rightarrow B_4C \qquad \text{... (4.67)}$$

Note that there was no release of CO or CO_2 in the above reactions. The average particle (agglomerate) size of the product was about 1μm and the specific surface area, 20 m^2/g.

Knudsen [629] used a laser-driven method of synthesis in which the reactants were BCl_3, H_2 and CH_4 (or C_2H_4). The interaction of the BCl_3 vapour stream with the laser caused the formation of a luminescent flame (temperature about 1500°C). The ratio of BCl_3/CH_4 was controlled for obtaining the proper stoichiometry (4 : 1) of the powder. The properties of the product were : loosely agglomerated, poorly crystallized B_4C, spheroidal shape, average diameter 34 nm, specific surface area 55 m^2/g, oxygen content 0.9 wt% and metal content extremely low (Fe 6 ppm).

4.2.6 Other Non-Oxide Powders

Zirconium Nitride

The major investigations on the preparation of zirconium nitride (ZrN) powders related to the reduction of ZrO_2 with magnesium in a nitrogen flow [630, 631]. The Mg/ZrO_2 molar ratio was kept at 5, and the reaction temperature varied in the range 500°–1100°C. The desired nitride formed above 600°C, but various by-products (e.g. MgO, Mg_3N_2) also co-existed, which had to be leached out to obtain relatively pure ZrN. The particles were reported to have a crystallite size of 16–18 nm and a specific surface area of 30–33 m^2/g.

In a second process [632], synthesis of ZrN powders was accomplished through the reaction between $ZrCl_4$ and N_2 gas in presence of Al and Mg metals as the reducing agents (the former more effective than the latter). Though the nitride powder formed above 800°C, crystallization of single-phase (0.1–0.3 μm) ZrN required 1050°C/1 h. The residual metal was removed by alkali wash.

Pyrolysis of complex precursors has been described as another route for obtaining zirconium nitride [194].

Titanium Nitride

Titanium nitride (TiN) powders are known to be prepared by two main methods :

(i) Direct nitridation of metal powder (or hydride) :

$$Ti + \tfrac{1}{2} N_2 \rightarrow TiN \qquad \text{... (4.68)}$$

(ii) Carbothermal reduction and nitridation of the oxide :

$$TiO_2 + 2C + \tfrac{1}{2} N_2 \rightarrow TiN + 2CO \qquad \text{... (4.69)}$$

An example of the direct nitridation comes from the investigations of Rubtsov et al. [633]. In this work, nitridation of electrolytic Ti (99.86% pure) particles of 40–80 μm size, and titanium hydride (about 4% H) was carried out in a steel reactor purged with argon and heated to 600°–1100°C after which pure N_2 was fed into it. Up to 17.60% N_2 could be incorporated into the precursor. The major problems were sintering of the Ti particles above 800°C and dehydration of TiH_2 above 700°C.

A glow discharge plasma was used by Singh and Doherty [634] for preparing TiN powders by nitridation of surface-cleaned Ti powders (coarse, 450–550 μm or fine, 100–150 μm) at 700°–1000°C. The fine powders were fully nitrided at 900°C/16 h.

Yoshimura et al. [635] have described an arc image heating system in which pure TiO_2 and graphite powders were mixed and pelletized before heating under N_2 for only 0.5 min to synthesize TiN. Conventional, longer-time heating at 850°– 1300°C has also been employed for the reaction.

Other techniques involve mixing of Ti particles with ammonium chloride, azides of Na, K or Ba, or TiN itself to obtain the nitride; titanium tetrachloride with metals like Na and Al has also been used as a starting material. Pyrolysis reaction involving $[(CH_3)_2N]_4Ti$ at 800°C is another route for preparing TiN. Narula [194] has recently made a brief discussion of these processes.

Zirconium Carbide

Zirconium carbide (ZrC) powders have been prepared [636] from the system ZrO_2-Mg-CH_4 by (i) direct synthesis from ZrO_2 + Mg and (ii) reduction of zirconia with Mg as the reducing agent; both the reactions occurred in a CH_4 gas flow.

The optimum conditions for direct synthesis were Mg/ZrO_2 molar ratio ≥ 4, reaction temperature $\geq 700°C$ and time 1 h; those for reduction with Mg were Mg/ZrO_2 molar ratio = 4, reduction time 1 h in argon flow and carbonization time 1 h in CH_4 flow at 800°C. Various intermediate products, however, formed during the reactions. The direct synthesis was found to be more effective.

The carbide has also been synthesized from $ZrCl_4$ and activated carbon [637] with magnesium as an effective reducing agent. With $Mg/ZrCl_4$ molar ratio = 3, ZrC and MgO were observed as the major phases, e.g. above 800°C. After acid-washing of the by-products, a single-phase ZrC (0.1–0.3 μm in size) powder was obtained.

Titanium Carbide

Reduction of titanium oxide through reaction with carbon is the oldest method for obtaining titanium carbide, TiC. The reduction has been reported [194] to start at 935°C and proceed to completion via the formation of suboxides. The high oxygen content of the product can be reduced by raising the temperature to 1600°–1700°C and providing a CO atmosphere. Other precursors are titanium tetrachloride, sulphide, alkoxide or the metal itself. Information is also available on a synthesis involving reactions of $TiCl_4$, CCl_4 and Mg at 700°–1300°C; $MgCl_2$ formed in the reaction sublimated at 800°–1200°C. Pure TiC was obtained through grinding and heating under vacuum at 1800°C [638].

Holt [628] has described a combustion technique where compacts of Ti and C were put together and heated by a tungsten coil under atmospheric pressure of argon, which was followed by immediate ignition. The reaction product, after dry-milling in a TiC-lined mill, had an average size of 6 μm.

Zirconium Diboride

The thermite method involving Mg (used in case of other non-oxide powders discussed earlier in this book) has been exploited also for zirconium diboride, ZrB_2.

Kobayashi et al. [639] used the combinations ZrO_2-B_2O_3-Mg and ZrO_2-amorphous B-Mg as the starting materials for accomplishing the following reactions :

$$ZrO_2 + B_2O_3 + 5Mg \rightarrow ZrB_2 + 5MgO \quad \text{... (4.70)}$$

$$ZrO_2 + 2B + 2Mg \rightarrow ZrB_2 + 2MgO \quad \text{... (4.71)}$$

The reduction of ZrO_2 (to α-zirconium) and B_2O_3 was found to occur simultaneously with the boriding action of α-zirconium and Mg. Phase-pure, fine particles of ZrB_2 were obtained at 1100°C (with B_2O_3) and above 700°C (with B). The particles were separated by deoxidized acidic solutions in a nitrogen atmosphere to remove MgO and MgB_2 as also to prevent oxidation of ZrB_2. The powder was found to contain up to 1 wt% Mg.

A mechanochemical synthesis of ZrB_2 has been reported recently [138].

Titanium Diboride

Titanium diboride, TiB_2 is the only stable compound in the binary system. Powders of this compound have been obtained commonly by the carbothermal reduction process. Details of the process and its economic aspects have been discussed by Kim and McMurtry [640].

A recent work of Millet and Hwang [138] on mechanochemical synthesis employed > 99% pure TiO_2 of particle size 0.2–0.3 μm and amorphous boron (92%) of particle size 0.2–0.4 μm. The Ti : Boron ratio was 1: 4. The mixture, when reacted in a high energy ball mill under vacuum for 25 h and heated at 1050°C, yielded TiB_2.

The diboride has also been obtained by the self-propagating high-temperature synthesis (SHS) involving the B_2O_3-TiO_2-Mg system. The reaction was characterized by pre-reaction steps related to the reduction of B_2O_3. The particles obtained were fine, uniformly distributed and agglomeration-free [641].

Another method of synthesis involved the reaction between the metal chlorides, controlled by sodium :

$$TiCl_4 + 2BCl_3 + 10\ Na \xrightarrow[110^\circ-160^\circ C]{n\text{-heptane}} \text{"}TiB_2\text{" (amorph)} + 10\ NaCl \quad ...\ (4.72)$$

A two-part glass reactor has been used by Ritter [642] for the reaction. The elemental halides and n-heptane, introduced in an upper flask, dropped slowly into a lower flask containing a dispersion of metallic sodium in toluene and n-heptane. After the reaction at 110°–160°C, the volatiles were removed by vacuum distillation (200°C) while NaCl was vacuum sublimated at 900°C. The black powder thus obtained was calcined at 1820°C in Ar to obtain well-crystallized TiB_2 (1–5 μm in size) with TiC and TiN as minor impurities. Following a similar but simple process, high-purity TiB_2 has been prepared [643].

4.2.7 Oxynitrides

Silicon Oxynitride

Silicon oxynitride, Si_2N_2O is generally not prepared in the powder form, but often obtained as an impurity phase during the synthesis of Si_3N_4. Information is therefore limited about powder synthesis; examples are given below.

Bolech et al. [644] describe a carbothermal reduction technique using silica sand

(5 μm, 99.8% SiO_2), carbon black (0.02 μm, 98% C) and nitrogen with ≤ 6 ppm O_2. The expected formation reaction was :

$$SiO_2 + \frac{3}{2}C + \frac{1}{2}N_2 \rightarrow \frac{1}{2}Si_2N_2O + \frac{3}{2}CO \qquad ... (4.73)$$

The other reactions to be considered with different SiO_2/C ratios were :

$$SiO_2 + 3C \rightarrow SiC + 2CO \qquad ... (4.74)$$

$$SiO_2 + C \rightarrow SiO + CO \qquad ... (4.75)$$

Bolech et al. [644] recognized two important precautions required to be taken during the oxynitride reaction :

(i) The formation and loss of the reduction product SiO (through the flow of oxygen) must be avoided to maintain the starting SiO_2/C ratio.

(ii) The temperature of reaction must be kept optimally low.

Considering this, a liquid-forming additive (SrO, 1–3 wt%) was used for assistance in the maximum possible conversion, which was achieved within about 1450°C (the lowest temperature of formation was 1350°C). Stoichiometric SiO_2/C ratio (= 2 : 3) produced the highest amount of the oxynitride, and no SiC.

In another investigation [645], a high surface area, spray-dried commercial pure silica sol was used for nitridation with the goal of obtaining the oxynitride. Dry ammonia was used as the reducing agent; the reaction took place at 1050°–1230°C for 20–47 h. At the higher temperatures, α-Si_3N_4 formed.

Silicon Aluminium Oxynitride

The elementary quaternary system Si-Al-O-N yields a family of solid solutions and interrelated structures with the overall family name SiAlON. The crystal chemistry and structures of these materials have been described in detail by Jack [e.g. 646].

SiAlON, also commonly spelt Sialon, is another example where powders are usually not prepared as the precursor for a ceramic. In case the preparation of a powder is the goal, carbothermal reduction of oxides of different sources is used as the principal technique. Mostaghaci et al. [647] have listed the different oxide starting materials : natural silica of different varieties, aluminium hydroxide gels and kaolin/metakaolin; aluminium powder has also been used in some cases. Some examples of powder preparation are given below.

In one investigation [648], a mixture was made of sedimentary kaolin, carbon black or coal and $Fe(NO_3).9H_2O$ or $FeCl_3.6H_2O$ (added as the source of Fe, a catalyst). The mixture was ground and pelletized and the pellets heated up to 1650°C in a nitrogen atmosphere. With 1.5–3% Fe, 90% SiAlON of a β'-Si_3N_4

structure could be synthesized at 1400°C/1 h, which increased to complete conversion at a slightly higher temperature. The product was milled to particles of about 1 μm size.

Yoshimatsu et al. [649] have reported a procedure in which natural kaolinite and carbon black were mixed and heated at 1350°–1500°C with up to 4 h soak in flowing nitrogen. With increase in the heating time, the carbon content decreased while the nitrogen content increased, indicating progressive nitridation. The β-SiAlON formed in the process was associated with other crystallized impurities like AlON and TiN (the titanium came from 1.7 wt% TiO_2 in the clay). At 1500°C, the associated impurities decreased in quantity, and a substantial amount of β-SiAlON was obtained.

Kaolin-carbon mixtures were also used for thermal treatment by Cho and Charles [650] in a carbothermal reaction under flowing nitrogen; the product was single-phase β′-SiAlON powder. It was shown that the kaolin/carbon ratio and silicon monoxide vaporization were of importance in controlling the phase composition of the product.

A combustion synthesis has been described by Zeng et al. [651] who used Si, Si_3N_4 and AlN powders as the starting materials. They were mixed in a methanol-filled polyethylene vessel, dried, sieved and packed in the container of a chamber pressurized (10 MPa) with N_2 gas of > 99.999% purity. A titanium powder pellet was placed on the top of the reaction mixture and ignited by a carbon ribbon heater to generate a temperature of 2,000°C. The reaction was :

$$Si + N_2 + Si_3N_4 + SiO_2 \text{ (from Si)} + AlN \rightarrow$$
$$Si_{6-z}Al_zO_zN_{8-z} \; (z = 0.3, 0.6) \qquad \ldots (4.76)$$

The obtained particles were 1–10 μm in size and needlelike with hexagonal facets. Silicon oxynitride was obtained as a by-product.

Aluminium Oxynitride

Aluminium oxynitride (ALON or AlON) is not a single compound, but can be obtained in various ratios of Al, O and N; as an associated feature, the oxynitride crystallizes in various different structures, e.g. polytypes based on wurtzite-type AlN structure or a spinel structure. The issue has been discussed in detail by Corbin [652].

Two major techniques are known to produce ALON powders [652–656] :

(i) Solid state reactions, (ii) Carbothermal reduction/nitridation of Al_2O_3.

Solid state reactions between Al_2O_3 and AlN at 1650°C have been reported to generate ALON powders [652]. For the carbothermal reduction/nitridation reaction [652–655] involving alumina, fine particles of the γ- and α-phases of Al_2O_3 and carbon black (up to about 7 wt%) have been mixed and heated up to 1850°C under flowing nitrogen. The heating was performed in two steps : in an initial step (1550°–1650°C), AlN was formed; it reacted at higher temperatures to produce the oxynitride, though both Al_2O_3 and AlN were generally present in the product. The aggolomerated powder thus obtained could be ball-milled to produce particles in the size range 0.5–5 μm [655].

A sol-gel technique has been used recently [656], in which the precursor was prepared from $AlCl_3.6H_2O$ and starch. Apparently the oxynitride was never phase-pure, but associated with Al_2O_3 and AlN.

References

1. G. Y. Onoda, Jr. and L. L. Hench, "Phy sical characterization terminology"; pp. 35–37 in *Ceramic Processing before Firing.* Edited by G. Y. Onoda, Jr. and L. L. Hench. John Wiley, New York, 1978.
2. M. Takahashi and S. Suzuki, "Compaction behaviour and mechanical characteristics of ceramic powders"; pp. 65–97 in *Handbook of Ceramics and Composites,* Vol. 1. Edited by N. P. Cheremisinoff. Marcel Dekker, New York, 1990.
3. H. Rumpf and H. Schubert, "Adhesion forces in agglomeration processes"; pp. 357–76 in *Ceramic Processing before Firing.* Edited by G. Y. Onoda, Jr. and L. L. Hench. John Wiley, New York, 1978.
4. A. Roosen and H. Hausner, "Techniques for agglomeration control during wet-chemical powder synthesis," *Adv. Ceram. Mater.*, **3**, 131–37 (1988).
5. R. G. Horn, "Surface forces and their action in ceramic materials," *J. Am. Ceram. Soc.*, **73**, 1117–35 (1990).
6. W. A. Ducker, Z. Xu, D. R. Clarke and J. N. Israelachvili,"Forces between alumina surfaces in salt solutions : non-DLVO forces and the implications for colloidal processing," *J. Am. Ceram. Soc.*, **77**, 437–43 (1994).
7. F. F. Lange and E. P. Luther, "Colloidal processing of structurally reliable Si_3N_4"; pp. 3–18 in *Tailoring of Mechanical Properties of Si_3N_4 Ceramics.* Edited by M. J. Hoffmann and G. Petzow. Kluwer Academic, Dordrecht, 1994.
8. M. N. Rahaman, *Ceramic Processing and Sintering*; p. 154. Marcel Dekker, New York, 1995.
9. P. C. Hiemenz, *Principles of Colloid and Surface Chemistry*; pp. 396–448. Marcel Dekker, New York, 1977.
10. H. C. Hamaker, "The London-van der Waals attractions between spherical particles," *Physica*, **4**, 1058–72 (1937).
11. D. Tabor, "Attractive surface forces"; pp. 23–46 in *Colloidal Dispersions*. Edited by J. D. Goodwin. Royal Society of Chemistry, London, 1982.
12. G. D. Parfitt, "Fundamental aspects of dispersion"; pp. 81–121 in *Dispersion of Powders in Liquids*. Edited by G. D. Parfitt. Elsevier, Amsterdam, 1969.
13. L. Lerot, F. Legrand and P. De Bruycker, "Chemical control in precipitation of spherical zirconia particles," *J. Mater. Sci.*, **26**, 2353–58 (1991).
14. T. Ikemoto, K.Uematsu, N. Mizutani and M. Kato, "Synthesis of monodispersed titania fine particles by hydrolysis of $Ti(OC_2H_5)_4$," *Yogyo-Kyokai-Shi*, **93**, 261–66 (1985). [in Japanese]
15. D. Burgard, C. Kropf, R. Nass and H. Schmidt, "Routes to deagglomerated nanopowders by chemical synthesis"; pp. 101–107 in *Mater. Res. Soc. Symp. Proc.*, **346**, *Better Ceramics Through Chemistry VI.* Edited by A. K. Cheetham, C. J. Brinker, M. L. Mecartney and C. Sanchez. Materials Research Society, Pittsburgh, PA, 1994.
16. R. Aveyard and D. A. Haydon, *An Introduction to the Principles of Surface Chemistry*; pp. 195–99. Cambridge University Press, Cambridge, 1973.

17. R. E. Johnson, Jr. and R. H. Dettre, "Wettability and contact angles"; pp. 85–153 in *Surface and Colloid Science,* **2.** Edited by E. Matijievic. Wiley-Interscience, New York, 1969.

18. E. A. Barringer and H. K. Bowen, "Ceramic powder processing," *Ceram. Eng. Sci. Proc.*, **5**, 285–97 (1984).

19. F. F. Lange, "Powder processing science and technology for increased reliability," *J. Am. Ceram. Soc.*, **72**, 3–15 (1989).

20. G. J. Vogt, D. S. Phillips and T. N. Taylor, "Plasma synthesis and characterization of ultrafine SiC"; pp. 203–15 in *Adv. Ceram.*, **21**, *Ceramic Powder Science.* Edited by G. L. Messing, K. S. Mazdiyasni, J. W. McCauley and R. A. Haber. American Ceramic Society, Westerville, OH, 1987.

21. N. Murakawa and M. Nakajima, "Agglomeration property of SiC powders prepared from fine SiO_2-C mixed powders," *J. Ceram. Soc. Japan*, **99**, 819–22 (1991). [in Japanese]

22. L. A. Harris, C. R. Kennedy, G. C.- T. Wei and F. P. Jeffers, "Microscopy of SiC powders synthesized by reacting colloidal silica and pitch," *J. Am. Ceram. Soc.*, **67**, C-121–C-124 (1984).

23. A. Cahn and J. L. Lynn, Jr., "Surfactants and detersive systems"; pp. 332–432 in *Kirk-Othmer Encyclopedia of Chemical Technology*, Vol. 22 (3rd ed). John Wiley, New York, 1983.

24. M. S. Newkirk, H. D. Lesher, D. R. White, C. R. Kennedy, A. W. Urquhart and T. D. Claar, "Preparation of Lanxide ceramic matrix composites : matrix formation by the directed oxidation of molten metals," *Ceram. Eng. Sci. Proc.*, **8**, 879–85 (1987).

25. K. T. Scott and R. Kingswell, "Thermal spraying"; pp. 217–43 in *Advanced Surface Coatings.* Edited by D. S. Rickerby and A. Matthews. Blackie, Glasgow, 1991.

26. H. Herman, "Plasma-sprayed coatings," *Scientific American, September,* 78–83 (1988).

27. K. T. Scott, "Plasma-sprayed ceramic coatings"; pp. 195–206 in *Brit. Ceram. Proc.*, **34**, *Ceramic Surfaces and Surface Treatments.* Edited by R. Morrell and M. G. Nicholas. The British Ceramic Society, Stoke-on-Trent, U. K., 1984.

28. H. J. Krause, "Ceramic pigments"; pp. 196–207 in *Process Mineralogy of Ceramic Materials.* Edited by W. Baumgart, A. C. Dunham and G. C. Amstutz. Ferdinand Enke Publishers, Stuttgart, 1984.

29. R. C. Schiek, "Pigments (inorganic)"; pp. 788–838 in *Kirk-Othmer Encyclopedia of Chemical Technology,* Vol. 17 (3rd ed.). John Wiley, New York, 1982.

30. R. H. Giller, "Colouring ceramic bodies"; pp. 191–96 in *Key Eng. Mater,* **53–55,** *Austceram '90.* Edited by P. J. Darragh and R. J. Stead. Trans Tech Publications, Switzerland, 1991.

31. U. S. Patent 2, 441, 407 (May, 1948).

32. G. Monros, M. C. Marti, J. Carda, M. A. Tena, V. Cantavella and J. Alarcon, "The encapsulating efficiency for vanadium of $ZrO_2.SiO_2$ gels prepared from alkoxides to give V-$ZrSiO_4$ blue coloured stains," *Brit. Ceram. Trans.*, **92**, 120–27 (1993).

33. R. Oheim, H. Paulus and C. Rüssel, "Preparation of praseodymium-doped $ZrSiO_4$ by a sol-gel route", *J. Mater. Sci. Lett.*, **10**, 1171–72 (1991).

34. Anon, *Ceram. Ind.*, **144** (1), 139 (1995).

35. K. J. Mills, "Ceramics and catalysis," *Ceram. Ind.*, **144** (6), 23–28 (1995).

36. J. R. Anderson, *Structure of Metallic Catalysts*; pp. 31–100. Academic Press, London, 1975.

37. L. Montanaro, A. Bachiorrini and A. Negro, "Deterioration of cordierite honeycomb structure for diesel emissions control," *J. Eur. Ceram. Soc.*, **13**, 129–34 (1994). Also references therein.

38. L. Coes, Jr., *Abrasives.* Springer-Verlag, Vienna, 1971.

39. W. G. Pinkstone, "Abrasives"; pp. 26–52 in *Kirk-Othmer Encyclopedia of Chemical Technology*, Vol. 1 (3rd ed.). John Wiley, New York, 1978.

40. M. G. Schwabel and P. E. Kendall, "Alumina abrasive grains produced by sol-gel technology," *Amer. Ceram. Soc. Bull.*, **70**, 1596–98 (1991).

41. U. S. Patent 4, 314, 827 (February, 1982).

42. U. S. Patent 4, 623, 364 (November, 1986).

43. U. S. Patent 4, 241, 000 (December, 1980).

44. W. M. Kenan, "Graphite, natural and synthetic particles," *Am. Ceram. Soc. Bull.*, **73**, 101–102 (1994).

45. R. S. Kalyoncu, "BN powder synthesis at low temperatures," *Ceram. Eng. Sci. Proc.*, **6**, 1356–64 (1985).

46. D. Fister, "AlN and BN powders for advanced applications," *Ceram. Eng. Sci. Proc.*, **6**, 1305–1312 (1985).

47. L. L. Hench, "Bioceramics : from concept to clinic," *J. Am. Ceram. Soc.*, **74**, 1487–1510 (1991).

48. A. Walter, "On the material and tribology of alumina-alumina coupling for hip joint prosthesis," *Clinic. Orthopaedics and Related Res.*, **282**, 31–48 (1992).

49. A. Ravaglioli, A. Krajewski, M. Dondi, A. Piancastelli, P. Ercolani, S. Tarlazzi and L. Fanti, "Chemical properties of various hydroxyapatite powders for industrial production," *Interceram*, **43**, 80–85 (1994).

50. E. A. Barringer and H. K. Bowen, "Formation, packing, and sintering of monodisperse TiO_2 powders," *J. Amer. Ceram. Soc.*, **65**, C-199–C-201 (1982).

51. Y. Suyama and A. Kato, "Effect of additives on the formation of TiO_2 particles by vapour phase reaction," *J. Am. Ceram. Soc.*, **68**, C-154–C-156 (1985).

52. T. Ogihara, H. Nakajima, T. Yanagawa, N. Ogata and K. Yoshida, "Preparation of monodisperse, spherical alumina powders from alkoxides," *J. Am. Ceram. Soc.*, **74**, 2263–69 (1991).

53. P. K. Das Poddar and M. Chaudhuri, "Natural minerals"; pp. 53–87 in *Hand Book of Ceramics*, Vol. 1. Edited by S. Kumar. Kumar and Associates, Calcutta, India, 1994.

54. J. M. Ayala, L. F. Verdeja, M. P. Garcia, M. A. Llavona and J. P. Sancho, "Production

of zirconia powders from the basic disintegration of zircon, and their characterization," *J. Mater. Sci*, **27**, 458–63 (1992).

55. T. E. Garner, Jr., "Zirconium raw materials for the refractory industries," *Ceram. Eng. Sci. Proc.*, **4**, 170–85 (1983).
56. Y. Shi, X. X. Huang and D. S. Yan, "Preparation and characterization of highly pure fine zircon powder," *J. Eur. Ceram. Soc.*, **13**, 113–19 (1994).
57. B. E. Leake, "Compilation of chemical analyses and physical constants of natural cordierites," *Am. Min.*, **45**, 282–98 (1960).
58. T. C. Labotka, J. J. Papike, D. T. Vaniman and G. B. Morey, "Petrology of contact metamorphosed argillite from the Rove formation, Gunflint trail, Minnesota," *Am. Min.*, **66**, 70–86 (1981).
59. B. Fegley, Jr., E. A. Barringer and H. K. Bowen, "Synthesis and characterization of monosized doped TiO_2 powders," *J. Am. Ceram. Soc.*, **67**, C-113–C-116 (1984).
60. J. H. Jean and T. A. Ring, "Nucleation and growth of monodisperse TiO_2 powders from alcohol solution in the presence of a sterically stabilizing surfactant"; pp. 11–33 in *Brit. Ceram. Proc.*, **38**, *Novel Ceramic Fabrication Processes and Applications*. Edited by R. W. Davidge. The Institute of Ceramics, Stoke-on-Trent, U. K., 1986.
61. K. Fujita and I. Kayama, "Synthesis of zinc oxide by the homogeneous precipitation method," *Yogyo-Kyokai-Shi*, **88**, 619–23 (1980). [in Japanese]
62. A. Nishida, K. Yoshida, H. Igarashi and W. Kobayashi, "Properties of ultrafine magnesium oxide powders prepared by vapour-phase oxidation"; pp. 271–77 in *Adv. Ceram.*, **21**, *Ceramic Powder Science*. Edited by G. L. Messing, K. S. Mazdiyasni, J. W. McCauley and R. A. Haber. American Ceramic Society, Westerville, OH, 1987.
63. I. J. McColm and N. J. Clark, *Forming, Shaping and Working of High-Performance Ceramics*; p. 142. Blackie, Glasgow, 1988.
64. I. B. Cutler, "Active powders"; pp. 21–29 in *Ceramic Processing before Firing*. Edited by G. Y. Onoda, Jr. and L. L. Hench. John Wiley, New York, 1978.
65. M. F. Yan, "Effects of physical, chemical, and kinetic factors on ceramic sintering"; pp. 635–69 in *Adv. Ceram.*, **21**, *Ceramic Powder Science*. Edited by G. L. Messing, K. S. Mazdiyasni, J. W. McCauley and R. A. Haber. American Ceramic Society, Westerville, OH, 1987.
66. C. Herring, "Effect of change of scale on sintering phenomena," *J. Appl. Phys.*, **21**, 301–303 (1950).
67. A. Kato., "Vapour phase synthesis of ultrafine non-oxide powders and their sintering behaviour," *Trans. Ind. Ceram. Soc.*, **51**, 6–16 (1992).
68. K. Kendall, "Influence of powder structure on processing and properties of advanced cermics," *Powder Technol.*, **58**, 151–61 (1989).
69. W. Stöber, A. Fink and E. Bohn, "Controlled growth of monodisperse silica spheres in the micron size range," *J. Colloid Interface Sci.*, **26**, 62–69 (1968).
70. M. Chatterjee and D. Ganguli, "Alkoxy-derived monodisperse silica microspheres : the role of solvents in synthesis," *Trans. Ind. Ceram. Soc.*, **45**, 95–99 (1986).

71. A. Kato, Y. Takeshita and Y. Katatae, "Preparation of spherical titania particles from inorganic precursor by homogeneous precipitation"; pp. 13–22 in *Mater. Res. Soc. Symp. Proc.*, **155**, *Processing Science of Advanced Ceramics*. Edited by I. A. Aksay, G. L. McVay and D. R. Ulrich. Materials Research Society, Pittsburgh, PA, 1989.

72. M. K. Akhtar, S. E. Pratsinis and S. V. R. Mastrangelo, "Dopants in vapour-phase synthesis of titania powders," *J. Am. Ceram. Soc.*, **75**, 3408–3416 (1992).

73. J. S. Chappell, J. D. Birchall and T. A. Ring, "The origin of defects arising in colloidal processing of sub-micron, monosize powders"; pp. 49–57 in *Brit. Ceram. Proc.*, **38**, *Novel Ceramic Fabrication Processes and Applications*. Edited by R. W. Davidge. The Institute of Ceramics, Stoke-on-Trent, U. K., 1986.

74. E. Barringer, N. Jubb, B. Fegley, R. L. Pober and H. K. Bowen, "Processing monosized powders"; pp. 315–33 in *Ultrastructure Processing of Ceramics, Glasses, and Composites*. Edited by L. L. Hench and D. R. Ulrich. John Wiley, New York, 1984.

75. L. Springer and M. F. Yan, "Sintering of TiO_2 from organometallic precursors"; pp. 464–75 in *Ultrastructure Processing of Ceramics, Glasses, and Composites*. Edited by L. L. Hench and D. R. Ulrich. John Wiley, New York, 1984.

76. M. F. Yan and W. W. Rhodes, "Low temperature sintering of TiO_2," *Mater. Sci. Eng.*, **61**, 59–66 (1983).

77. R. B. Heady and J. W. Cahn, "Analysis of capillary forces in liquid-phase sintering of spherical particles," *Metallurgical Trans.*, **1**, 185–89 (1970).

78. J. W. Cahn and R. B. Heady, "Analysis of capillary forces in liquid-phase sintering of jagged particles," *J. Am. Ceram. Soc.*, **53**, 406–409 (1970).

79. E. Matijevic, "Monodispersed metal (hydrous) oxides–a fascinating field of colloid science," *Acc. Chem. Res.*, **14**, 22–29 (1981).

80. G. Wötting and G. Ziegler, "Powder characteristics and sintering behaviour of Si_3N_4–powders," *Powder Metallurgy Int.*, **18**, 25–32 (1986).

81. F. F. Lange, "Fracture toughness of Si_3N_4 as a function of the initial a-phase content," *J. Am. Ceram. Soc.*, **62**, 428–30 (1979).

82. M. J. Hoffmann, "Analysis of microstructural development and mechanical properties of Si_3N_4 ceramics"; pp. 59–72 in *Tailoring of Mechanical Properties of Si_3N_4 Ceramics*. Edited by M. J. Hoffmann and G. Petzow. Kluwer Academic, Dordrecht, 1994.

83. K. Fujita, K. Murata, T. Nakazawa and I. Kayama, "Crystal shapes of zinc oxide prepared by the homogeneous precipitation method," *Yogyo-Kyokai-Shi*, **92**, 227–30 (1984).

84. Y. Sakka, K. Halada and E. Ozawa, "Synthesis of ZnO particles by the homogeneous precipitation method"; pp. 31–38 in *Ceram. Trans.*, **1**, *Ceramic Powder Science II*. Edited by G. L. Messing, E. R. Fuller, Jr. and H. Hausner. American Ceramic Society, Westerville, OH, 1988.

85. M. E. V. Costa and J. L. Baptista, "Characteristics of zinc oxide powders precipitated in the presence of alcohols and amines," *J. Eur. Ceram. Soc.*, **11**, 275–81 (1993).

86. U. S. Patent 5, 171, 364 (1992).

87. R. J. Lauf and W. D. Bond, "Fabrication of high-field zinc oxide varistors by sol-gel processing," *Am. Ceram. Soc. Bull.*, **63**, 278–81 (1984).

88. G. MacZura, K. J. Moody, E. M. Anderson and M. K. Kunka, "Alumina," *Am. Ceram. Soc. Bull.*, **72** (6), 76–78 (1993).

89. E. S. Martin and M. L. Weaver, "Synthesis and properties of high-purity alumina," *Am. Ceram. Soc. Bull.*, **72** (7), 71–77 (1993).

90. E. M. Anderson, "Developments in powder processing methods," *Ceram. Ind.*, **144** (1), 33–37 (1995).

91. J. S. Reed, *Principles of Ceramics Processing* (2nd ed.); pp. 57 and 71. John Wiley, New York, 1995.

92. A. W. Weimer, G. A. Cochran, G. A. Eisman, J. P. Henley, B. D. Hook, L. K. Mills, T. A. Guiton, A. K. Knudsen, N. R. Nicholas, J. E. Volmering and W. G. Moore, "Rapid process for manufacturing aluminium nitride powder," *J. Am. Ceram. Soc.*, **77**, 3–18 (1994).

93. E. Ponthieu, P. Grange, B. Delmon, L. Lonnoy, L. Leclercq, R. Bechara and J. Grimblot, "Proposal of a composition model for commercial AlN powders," *J. Eur. Ceram. Soc.*, **8**, 233–41 (1991).

94. S. Sahu, S. Kavecky and J. Szepvolgyi, "Preparation of fine amorphous silicon nitride powder in the system SiH_4-Ar-NH_3," *J. Eur. Ceram. Soc.*, **15**, 1071–77 (1995).

95. S. G. Malghan, "Silicon nitride powders and their processing"; pp. 243–63 in *Key Eng. Mater.*, **56–57**, *Advanced Ceramics*. Edited by C. Ganguly, S. K. Roy and P. R. Roy. Trans Tech Publications, Zurich, 1991.

96. T. Hayashi, H. Ushida, H. Saito and S. Hirano, "Effects of NaF and NH_3 on preparation of Si_3N_4 powders from SiO_2," *Yogyo-Kyokai-Shi*, **95**, 278–83 (1987). [in Japanese]

97. M. H. Lewis, G. Leng-Ward and C. Jasper, "Sintering additive chemistry in controlling microstructure and properties of nitride ceramics"; pp. 1019–33 in *Ceram. Trans.*, **1**, *Ceramic Powder Science II*. Edited by G. L. Messing, E. R. Fuller, Jr. and H. Hausner, American Ceramic Society, Westerville, OH, 1988.

98. K. Okada, K. Fukuyama and Y. Kameshima, "Characterization of surface-oxidized phase in silicon nitride and silicon oxynitride powders by X-ray photoelectron spectroscopy," *J. Am. Ceram. Soc.*, **78**, 2021–26 (1995).

99. Japanese Patent 5891, 009 (May, 1983).

100. S. Natansohn, A. E. Pasto and W. J. Rourke, "Effect of powder surface modifications on the properties of silicon nitride ceramics," *J. Am. Ceram. Soc.*, **76**, 2273–84 (1993).

101. R. N. Singh, "High-temperature mechanical properties of a uniaxially reinforced zircon-silicon carbide composite," *J. Am. Ceram. Soc.*, **73**, 2399–2406 (1990).

102. T. Mori, H. Hoshino, H. Yamamura, H. Kobayashi and T. Mitamura, "Mechanical properties of high purity sintered $ZrSiO_4$," *J. Ceram. Soc. Japan*, **98**, 1017–22 (1990). [in Japanese]

103. Japanese Patent 1–226709 (1989).

104. K. Uchiyama, T. Ogihara, T. Ikemoto, N. Mizutani and M. Kato, "Preparation of monodispersed Y-doped ZrO_2 powders," *J. Mater. Sci.*, **22**, 4343–47 (1987).

105. H. Ishizawa, O. Sakurai, N. Mizutani and M. Kato, "Homogeneous Y_2O_3-stabilized ZrO_2 powder by spray pyrolysis method," *Am. Ceram. Soc. Bull.*, **65**, 1399–1404 (1986).

106. K. Takigawa, K. Nonaka, K. Okada and N. Otsuka, "Preparation of PLZT fine powders by the spray pyrolysis nethod," *Brit. Ceram. Trans. J.*, **89**, 82–86 (1990).

107. D. C. Agrawal, S. B. Majumder and G. V. Ramani, "Coated powders for ceramics and composites"; pp. 165–82 in *Key Eng. Mater.*, **56–57**, *Advanced Ceramics*. Edited by C. Ganguly, S. K. Roy and P. R. Roy. Trans Tech Publications, Zurich, 1991.

108. R. Metselaar, R. Reenis, M. Chen, H. Gorter and H. T. Hintzen, "Surface characterization of chemically treated aluminium nitride powders," *J. Eur. Ceram. Soc., 15*, 1079–85 (1995).

109. T. Reetz, B. Mönch and M. Saupe, "Aluminium nitride hydrolysis," *Ceram. Forum Int. / Ber. DKG*, **69**, 464–65 (1992) .

110. P. Bowen, J. G. Highfield, A. Mocellin and T. A. Ring, "Degradation of aluminium nitride powder in aqueous environment," *J. Am. Ceram. Soc.*, **73**, 724–28 (1990).

111. M. Egashira, Y. Shimizu, Y. Takao, R. Yamaguchi and Y. Ishikawa, "Effect of carboxylic acid adsorption on the hydrolysis and sintered properties of aluminium nitride powder," *J. Am. Ceram. Soc.*, **77**, 1793–98 (1994).

112. D. Hotza and P. Greil, "Hydrophobicity of aluminium nitride," *Fortschrittsberichte Deutsch. Keram. Gesell.*, **9**, 107–121 (1994). [in German]

113. H. Schmidt, G. Nabert, G. Ziegler and H. Goretzki, "Characterization and surface chemistry of uncoated and coated silicon nitride powders," *J. Eur. Ceram. Soc.*,**15**, 667–74 (1995).

114. B. Djuricic, I. J. Davies, S. Pickering, D. McGarry, E. Bullock, M. Verwerft, P. M. Bronsveld and J. Th. M. De Hosson, "Study of particle coatings for the design of intergranular phases in engineering ceramics," *Silic. Ind.*, **60**, 203–210 (1995).

115. C.-M. Wang and F. L. Riley, "Alumina-coating of silicon nitride powder," *J. Eur. Ceram. Soc.*, **10**, 83–93 (1992).

116. W.-H. Shih and L.-L. Pwu, "Rheology of aqueous boehmite-coated silicon nitride suspensions and gels," *J. Mater. Res.*, **10**, 2808–2816 (1995).

117. K. Kishi, S. Umebayashi, R. Pompe and M. Persson, "Interaction of aluminium-iso-propoxide solution and Si_3N_4 powder," *J. Ceram. Soc. Japan*, **96**, 698–701 (1988). [in Japanese]

118. E. P. Luther, F. F. Lange and D. S. Pearson, " 'Alumina' surface modification of silicon nitride for colloidal processing," *J. Am. Ceram. Soc.*, **78**, 2009–2014 (1995).

119. S. Baik and R. Raj, "Effect of silicon activity on liquid-phase sintering of nitrogen ceramics," *J. Am. Ceram. Soc.*, **68**, C-124–C-126 (1985).

120. M. Kulig, W. Oroschin and P. Greil "Sol-gel coating of silicon nitride with Mg-Al oxide sintering aid," *J. Eur. Ceram. Soc.*, **5**, 209–217 (1989).

121. J.-S. Kim, H. Schubert and G. Petzow, "Sintering of Si_3N_4 with Y_2O_3 and Al_2O_3 added by coprecipitation," *J. Eur. Ceram. Soc.*, **5**, 311–19 (1989).

122. T.-I. Mah, K. S. Mazdiyasni and R. Ruh, "Characterization and properties of hot-pressed Si_3N_4 with alkoxy-derived CeO_2 or Y_2O_3 as sintering aids," *Am. Ceram. Soc. Bull.*, **58**, 840–44 (1979).

123. T. M. Shaw and B. A. Pethica, "Preparation and sintering of homogeneous silicon nitride green compacts," *J. Am. Ceram. Soc.*, **69**, 88–93 (1986).

124. L. Bergström and E. Bostedt, "Surface chemistry of silicon nitride powders, electrokinetic behaviour and ESCA studies," *Colloids and Surfaces*, **49**, 183–97 (1990).

125. W. L. McCabe and J. C. Smith, *Unit Operations of Chemical Engineering* (3rd ed.); pp. 818–51. McGraw-Hill, Tokyo, 1976.

126. I. J. McColm and N. J. Clark, *Forming, Shaping and Working of High-Performance Ceramics*; p. 104. Blackie, Glasgow, 1988.

127. A. Szegvari, "The fine grinding of ceramics with attritors," *Interceram*, **43**, 97–98 (1994).

128. Swiss Patent 132086 (January 1928).

129. A. Russell, "Fine grinding – a review," *Ind. Min.*, 58–70, April 1989.

130. W. Jander, "Reactions in solid state at high temperatures : I," *Z. Anorg. Allgem. Chem.*, **163**, 1–30 (1927). [in German]

131. S. F. Hulbert, "Models for solid-state reactions in powdered compacts : a review," *J. Brit. Ceram. Soc.*, **6**, 11–20 (1969).

132. R. E. Carter, "Kinetic model for solid-state reactions," *J. Chem. Phys.*, **34**, 2010–15 (1961).

133. R. E. Carter, "Addendum : kinetic model for solid-state reactions," *J. Chem. Phys.*, **35**, 1137–38 (1961).

134. M. S. El-Eskandarany, K. Sumiyama and K. Suzuki, "Mechanical solid state reaction for synthesis of b-SiC powders," *J. Mater. Res.*, **10**, 659–67 (1995).

135. P. Matteazzi, G. Le Caer and E. Bauer-Grosse, "Synthesis of advanced ceramics by high energy milling," *Key Eng. Mater.*, **53–55**, 451–56 (1991).

136. P. Matteazzi and G. Le Caer, "Room-temperature mechanosynthesis of carbides by grinding of elemental powders," *J. Am. Ceram. Soc.*, **74**, 1382–90 (1991).

137. M. N. Rahaman, *Ceramic Processing and Sintering*; pp. 50–54. Marcel Dekker, New York, 1995.

138. P. Millet and T. Hwang, "Preparation of TiB_2 and ZrB_2. Influence of a mechanochemical treatment on the borothermic reduction of titania and zirconia," *J. Mater. Sci*, **31**, 351–55 (1996).

139. L. M. Sheppard, "Manufacturing ceramics with microwaves : the potential for economical production," *Am. Ceram. Soc. Bull.*, **67**, 1656–61 (1988).

140. W. H. Sutton, "Microwave processing of ceramic materials," *Am. Ceram. Soc. Bull.*, **68**, 376–86 (1989).

141. D. M. P. Mingos, "Microwave syntheses of materials and their precursors"; pp. 717–25 in *Chemical Processing of Advanced Materials*. Edited by L. L. Hench and J. K. West. John Wiley, New York, 1992.

142. P. D. Ramesh, B. Vaidhyanathan, M. Ganguli and K. J. Rao, "Synthesis of b-SiC powder by use of microwave radiation," *J. Mater. Res.*, **9**, 3025–27 (1994).

143. K. J. Rao and P. D. Ramesh, "Use of microwaves for the synthesis and processing of materials," *Bull. Mater. Sci*, **18**, 447–65 (1995).

144. J. W. Mullin, *Industrial Crystallization*. Plenum Press, New York, 1976.

145. A. G. Jones, "Particle formation during agglomerative precipitation processes"; pp. 61–94 in *Controlled Particle, Droplet and Bubble Formation*. Edited by D. J. Wedlock. Butterworth-Heinemann, Oxford, 1994.

146. C. F. Baes, Jr. and R. E. Mesmer, *The Hydrolysis of Cations*. John Wiley, New York, 1976.

147. Y. K. Hong, Y. J. Paig, D. J. Agresti and T. D. Shelfer, "Synthesis and characterization of modified barium ferrite particles," *J. Appl. Phys.*, **61**, 3872–74 (1987).

148. B. H. Davis, "Effect of pH on crystal phase of ZrO_2 precipitated from solution and calcined at 600°C ," *J. Am. Ceram. Soc.*, **67**, C-168 (1984).

149. D. W. Johnson, Jr. and P. K. Gallagher , "Reactive powders from solution"; pp. 125–39 in *Ceramic Processing before Firing*. Edited by G. Y. Onoda, Jr. and L. L. Hench. John Wiley, New York, 1978.

150. M. Murata, K. Wakino, K. Tanaka and Y. Hamakawa, "Chemical preparation of PLZT powder from aqueous solution," *Mater. Res. Bull.*, **11**, 323–28 (1976).

151. E. Matijevic, "Formation of monodisperse inorganic particulates"; pp. 39–59 in *Controlled Particle, Droplet and Bubble Formation*. Edited by D. J. Wedlock. Butterworth-Heinemann, Oxford, 1994.

152. H. Nagai, S. Hokazono and A. Kato, "Synthesis of aluminium hydroxide by a homogeneous precipitation method I—effect of additives on the morphology of aluminium hydroxide," *Brit. Ceram. Trans. J.*, **90**, 44–48 (1991).

153. S. Somiya, K. Ioku and M. Yoshimura, "Hydrothermal synthesis and characterization of fine apatite crystals"; pp. 371–78 in *Mater. Sci. Forum,* **34–36**, *Austceram 88*. Edited by C. C. Sorrell and B. Ben-Nissan. Trans Tech Publications, Switzerland, 1988.

154. A. Somiya, K. Hishinuma and T. Akiba, "A new materials processing – hydrothermal processing," *Bull. Mater. Sci.*, **18**, 811–18 (1995).

155. M. Yoshimura and S. Somiya, "Fine zirconia powders by hydrothermal processing," *Rept. Res. Lab. Eng. Mater., Tokyo Inst. Technol.*, **9**, 53–64 (1984).

156. P. Reynen, H. Bastius, B. Pavlovski and D. von Mallinckrodt, "Production of high-purity zirconia by hydrothermal decomposition of zircon"; pp. 464–75 in *Adv. Ceram.*, **3**, *Science and Technology of Zirconia*. Edited by A. H. Heuer and L. W. Hobbs. American Ceramic Society, Columbus, OH, 1981.

157. J. L. Woodhead, "Sol-gel processes to ceramic particles using inorganic precursors," *J. Mater. Educ.*, **6**, 888–925 (1984).

158. C. J. Brinker and G. W. Scherer, *Sol-Gel Science. Academic Press, Boston, 1990.*

159. U. S. Patent 3, 330, 697 (July, 1967).

160. K. D. Budd and D. A. Payne, "Preparation of strontium titanate ceramics and internal boundary layer capacitors by the Pechini method"; pp. 239–44 in *Mater. Res. Soc. Symp. Proc.*, **32**, *Better Ceramics Through Chemistry.* Edited by C. J. Brinker, D. E. Clark and D. R. Ulrich. Elsevier, New York, 1984.

161. K. S. Mazdiyasni, "Powder synthesis from metal-organic precursors," *Ceram. Int.*, **8**, 42–56 (1982).

162. R. W. Hartel and K. A. Berglund, "Precipitation kinetics of the titanium isopropoxide hydrolysis reaction"; pp. 633–40 in *Mater. Res. Soc. Symp. Proc.*, **73**, *Better Ceramics Through Chemistry II.* Edited by C. J. Brinker, D. E. Clark and D. R. Ulrich. Materials Research Society, Pittsburgh, PA, 1986.

163. K. S. Mazdiyasni, R. T. Dolloff and J. S. Smith II, "Preparation of high-purity submicron barium titanate powders," *J. Am. Ceram. Soc.*, **52**, 523–26 (1969).

164. A. B. Hardy, G. Gowda, T. J. McMahon, R. E. Riman, W. E. Rhine and H. K. Bowen, "Preparation of oxide powders"; pp. 407–428 in *Ultrastructure Processing of Advanced Ceramics.* Edited by J. D. Mackenzie and D. R. Ulrich. John Wiley, New York, 1988.

165. T. Ogihara, H. Kaneko, N. Mizutani and M. Kato, "Preparation of monodispersed lead zirconate-titanate fine particles," *J. Mater. Sci. Lett.*, **7**, 867–69 (1988).

166. E. Dickinson, "Emulsions and droplet size control"; pp. 191–216 in *Controlled Particle, Droplet and Bubble Formation.* Edited by D. J. Wedlock. Butterworth-Heinemann, Oxford, 1994.

167. M. Chatterjee, B. Siladitya and D. Ganguli, "Chromia microspheres by the sol-gel technique," *Mater. Lett.*, **25**, 261–63 (1995).

168. W. Black, "Surface-active compounds"; pp. 201–239 in *Dispersion of Powders in Liquids.* Edited by G. D. Parfitt. Elsevier, Amsterdam, 1969.

169. M. Bourrel and R. S. Schechter, *Microemulsions and Related Systems*; p. 25. Marcel Dekker, New York, 1988.

170. J. Thomson, Jr., "Chemical preparation of PLZT powders from aqueous nitrate solutions," *Am. Ceram. Soc. Bull.*, **53**, 421–24, 433 (1974).

171. L. M. Sheppard, "Trends in powder processing equipments," *Am. Ceram. Soc. Bull.*, **72** (5), 28–37 (1993).

172. S. J. Lukasiewicz, "Spray-drying ceramic powders," *J. Am. Ceram. Soc.*, *72*, 617–24 (1989).

173. M. Chatterjee, J. Ray and D. Ganguli, "Spray-drying of hydrated zirconia slurries : a laboratory study," *Brit. Ceram. Trans. J.*, **91**, 159–61 (1992).

174. G. L. Messing, S.-C. Zhang and G. V. Jayanthi, "Ceramic powder synthesis by spray pyrolysis," *J. Am. Ceram. Soc.*, **76**, 2707–2726 (1993).

175. D. M. Roy, R. R. Neurgaonkar, T. P. O'Holleran and R. Roy, "Preparation of fine oxide powders by evaporative decomposition of solutions," *Am. Ceram. Soc. Bull.*, **56**, 1023–24 (1977).

176. M. J. Ruthner, "Industrial production of multicomponent ceramic powders (metal oxides) by means of the spray roasting technique"; pp. 515–31 in *Ceramic Powders*. Edited by P. Vincenzini. Elsevier, Amsterdam, 1983.

177. N. Mizutani and T. Q. Liu, "Synthesis of spherical Si_3N_4 powders by spray pyrolysis of polysilazane"; pp. 59–73 in *Ceram. Trans.*, **12**, *Ceramic Powder Science III*. Edited by G. L. Messing, S. Hirano and H. Hausner. American Ceramic Society, Westerville, OH, 1990.

178. F. J. Schnettler, F. R. Monforte and W. W. Rhodes, "A cryochemical method for preparing ceramic materials"; pp. 79–90 in *Sci. Ceram.*, **4**, Edited by G. H. Stewart. British Ceramic Society, Stoke-on-Trent, U. K., 1968.

179. F. Dogan and H. Hausner, "The role of freeze-drying in ceramic powder processing," pp. 127–34 in *Ceram. Trans.*, **1**, *Ceramic Powder Science II*. Edited by G. L. Messing, E. R. Fuller, Jr. and H. Hausner. American Ceramic Society, Westerville, OH, 1988.

180. H. K. Pulker, *Coatings on Glass*; Chapter 6. Elsevier, Amsterdam, 1984.

181. N. Ichinose, Y. Ozaki and S. Kashu, *Superfine Particle Technology*; Chapter 4. Springer-Verlag, London, 1992.

182. D. W. Johnson, Jr., "Innovations in ceramic powder preparation"; pp. 3–19 in *Adv. Ceram.*, **21**, *Ceramic Powder Science*. Edited by G. L. Messing, K. S. Mazdiyasni, J. W. McCauley and R. A. Haber, American Ceramic Society, Westerville, OH, 1987.

183. R. W. Sigel, S. Ramasamy, H. Hahn, L. Zongquan and L. Ting, "Synthesis, characterization and properties of nanophase TiO_2," *J. Mater. Res.*, **3**, 1367–72 (1988).

184. J. S. Haggerty, "Growth of precisely controlled powders from laser heated gases"; pp. 353–66 in *Ultrastructure Processing of Ceramics, Glasses, and Composites*. Edited by L. L. Hench and D. R. Ulrich. John Wiley, New York, 1984.

185. Y. Suyama, R. M. Marra, J. S. Haggerty and H. K. Bowen, "Synthesis of ultrafine SiC powders by laser-driven gas phase reactions," *Am. Ceram. Soc. Bull.*, **64**, 1356–59 (1985).

186. R. A. Marra and J. S. Haggerty, "Synthesis and characteristics of ceramic powders made from laser-heated gases," *Ceram. Eng. Sci. Proc.*, **3**, 3–19 (1982).

187. K. S. Mazdiyasni and C. M. Cooke, "Synthesis, characterization, and consolidation of Si_3N_4 obtained from ammonolysis of $SiCl_4$," *J. Am. Ceram. Soc.*, **56**, 628–33 (1973).

188. T. Yamada, "Preparation and evaluation of sinterable silicon nitride powder by imide decomposition method," *Am. Ceram. Soc. Bull*, **72** (5), 99–106, 119 (1993).

189. M. Takase and T. Sata, "Formation mechanism of ultrafine alumina particle from NH_3 gas–$AlCl_3$ solution," *Yogyo–Kyokai–Shi*, **87**, 495–500 (1979).

190. M. N. Rahaman, *Ceramic Processing and Sintering*; p. 82. Marcel Dekker, New York, 1995.

191. J. M. Haussonne, J. Lostec, J. P. Bertot, L. Lostec and S. Sadou, "A new synthesis process for AlN," *Am. Ceram. Soc. Bull.*, **72** (5), 84–90 (1993).

192. Advanced Ceramic Powders. Product booklet of H. C. Starck GmbH & Co., Germany (1996).

193. W. S. Clabaugh, E. M. Swiggard and R. Gilchrist, "Preparation of barium titanyl oxalate tetrahydrate for conversion to barium titanate of high purity," *J. Res. National Bureau of Standards*, **56**, 289–91 (1956).

194. C. K. Narula, *Ceramic Precursor Technology and its Applications*. Marcel Dekker, New York, 1995.

195. D. Segal, *Chemical Synthesis of Advanced Ceramic Materials*; pp. 96–113. Cambridge University Press, Cambridge, 1989.

196. G. T. Burns and G. Chandra, "Pyrolysis of preceramic polymers in ammonia : preparation of silicon nitride powders," *J. Am. Ceram. Soc.*, **72**, 333–37 (1989).

197. M. M. Seibold and C. Rüssel, "Thermal conversion of preceramic polyiminoalane precursors to aluminium nitride : characterization of pyrolysis products," *J. Am. Ceram. Soc.*, **72**, 1503–1505 (1989).

198. British Patent 10, 093 (1887).

199. W. H. Gitzen, *Alumina as a Ceramic Material*. American Ceramic Society, Ohio, 1970.

200. J. C. Southern, "Ceramic aluminas"; pp. 1–11 in *Brit. Ceram. Proc.*, **47**, *Fine Ceramic Powders*. Edited by R. Freer and J. L. Woodhead. The Institute of Ceramics, Stoke-on-Trent, U. K., 1991.

201. D. P. F. de Souza and M. F. de Souza, "Alumina purification by carbothermal reduction," *J. Mater. Chem.*, **6**, 233–38 (1996).

202. R. Brace and E. Matijevic, "Aluminium hydrous oxide sols – I. Spherical particles of narrow size distribution," *J. Inorg. Nucl. Chem.*, **35**, 3691–3705 (1973).

203. K. Fujita, S. Konno and I. Kayama, "Particle size of alumina hydrate gel prepared by the homogeneous precipitation method," *Yogyo-Kyokai-Shi*, **94**, 601–603 (1986). [in Japanese]

204. H. Nagai, Y. Oshima, K. Hirano and A. Kato, "Sintering behaviour of aluminium oxides derived from aluminium hydroxides with various morphologies," *Brit. Ceram. Trans.*, **92**, 114–19 (1993).

205. B. C. Cornilsen and J. S. Reed, "Homogeneous precipitation of basic aluminium salts as precursors for alumina," *Am. Ceram. Soc. Bull.*, **58**, 1199 (1979).

206. M. D. Sacks, T. Y. Tseng and S. Y. Lee, "Thermal decomposition of spherical hydrated basic aluminium sulphate," *Am. Ceram. Soc. Bull.*, **63**, 301–310 (1984).

207. J. E. Blendell, H. K. Bowen and R. L. Coble, "High purity alumina by controlled precipitation from aluminium sulphate solutions," *Am. Ceram. Soc. Bull.*, **63**, 797–802 (1984).

208. D. W. Johnson and F. J. Schnettler, "Characterization of freeze-dried Al_2O_3 and Fe_2O_3," *J. Am. Ceram. Soc.*, **53**, 440–44 (1970).

209. L. Wang and I. K. Lloyd, "Sinterability of calcined freeze-dried alumina powders," *J. Am. Ceram. Soc.*, **74**, 2934–36 (1991).

210. K. Okada, A. Tanaka, S. Hayashi and N. Otsuka, "Preparation of Al_2O_3 powders from various aluminium salts by the spray pyrolysis method," *J. Mater. Sci. Lett.*, **12**, 854–57 (1993).

211. K. Okada, A. Tanaka, S. Hayashi, K. Daimon and N. Otsuka, "Porous alumina ceramics by spray-pyrolyzed powder from aluminium sulphate and aluminium nitrate solutions," *J. Mater. Res.*, **9**, 1709–1713 (1994).

212. R. E. Jaeger and T. J. Miller, "Preparation of ceramic oxide powders by liquid drying," *Am. Ceram. Soc. Bull.*, **53**, 855–59 (1974).

213. D. W. Johnson, Jr. and P. K. Gallagher, "Kinetics of the decomposition of freeze-dried aluminium sulphate and ammonium aluminium sulphate," *J. Am. Ceram. Soc.*, **54**, 461–65 (1971).

214. E. Kato, K. Daimon, M. Nanbu, "Decomposition of two aluminium sulphates and characterization of the resultant aluminas," *J. Am. Ceram. Soc.*, **64**, 436–43 (1981).

215. K. Hayashi, S. Toyoda, K. Nakashima and K. Morinaga, "Optimum synthetic conditions of ammonium aluminium carbonate hydroxide (AACH) as starting material for a-alumina fine powders," *J. Ceram. Soc. Japan,* **98**, 444–49 (1990). [in Japanese]

216. J. L. Henry and H. J. Kelly, "Preparation and properties of ultrafine high-purity alumina," *J. Am. Ceram. Soc.*, **48**, 217–18 (1965).

217. F. W. Dynys and J. W. Halloran, "Alpha alumina formation in alum-derived gamma alumina," *J. Am. Ceram. Soc.*, **65**, 442–48 (1982).

218. Y. Sarikaya and M. Akinc, "Preparation of alumina microshells by the emulsion evaporation technique," *Ceram. Int.*, **14**, 239–44 (1988).

219. J. Ray, M. Chatterjee and D. Ganguli, "Sol-gel derived alumina microspheres," *J. Mater. Sci. Lett.*, **12**, 1755–57 (1993).

220. A. J. Fanelli and J. V. Burlew, "Preparation of fine alumina powder in alcohol," *J. Am. Ceram. Soc.*, **69**, C-174–C-175 (1986).

221. H.-L. Chang and I. K. Lloyd, "Alumina prepared from freeze-dried aluminium isopropoxide," *J. Am. Ceram. Soc.*, **76**, 1357–60 (1993).

222. T. Hirayama, "High-temperature characteristics of transition Al_2O_3 powder with ultrafine spherical particles," *J. Am. Ceram. Soc.*, **70**, C-122–C-124 (1987).

223. D. A. Everest, I. G. Sayce and B. Selton, "Preparation of ultrafine alumina powders by plasma evaporation," *J. Mater. Sci.*, **6**, 218–24 (1971).

224. A. T. Liu and P. Kleinschmit, "Production of fumed oxides by flame hydrolysis"; pp. 1–10 in *Brit. Ceram. Proc.*, **38**, *Novel Ceramic Fabrication Processes and Applications.* Edited by R. W. Davidge. The Institute of Ceramics, Stoke-on-Trent, U. K., 1986.

225. K. Kamata, T. Mochizuki, S. Matsumoto, A. Yamada and K. Miyokawa, "Preparation of submicrometer Al_2O_3 powder by gas-phase oxidation of tris (acetylacetonato) aluminium (III)," *J. Am. Ceram. Soc.*, **68**, C-193–C-194 (1985).

226. J. J. Kingsley and K. C. Patil, "A novel combustion process for the synthesis of fine particle a-alumina and related oxide materials," *Mater. Lett.*, **6**, 427–32 (1988).

227. M. Rühle and A. H. Heuer, "Phase transformations in ZrO_2-containing ceramics : II, the martensitic reaction in t-ZrO_2"; pp. 14–32 in *Adv. Ceram.*, **12**, *Science and Technology of Zirconia II.* Edited by N. Claussen, M. Rühle and A. H. Heuer. American Ceramic Society, Columbus, OH, 1984.

228. R. Stevens, *Zirconia and Zirconia Ceramics*. Magnesium Elektron Ltd., U. K., 1986.

229. M. R. Houchin, D. H. Jenkins and H. N. Sinha, "Production of high-purity zirconia from zircon," *Am. Ceram. Soc. Bull.*, **69**, 1706–1710 (1990).

230. H. S. Choi, "Preparation of pure zirconyl compounds from zircon caustic frit," *Canad. Mining and Metallurgical Bull.*, **68**, 193–98 (1965).

231. P. R. Menon, J. M. Juneja and T. S. Krishnan, "Decomposition of zircon by soda ash sintering process," *Am. Ceram. Soc. Bull.*, **59**, 635–36, 645 (1980).

232. R. H. Nielsen and R. L. Govro, "Zirconium purification, using a basic sulphate precipitation"; pp. 1–14 in U. S. Bureau of Mines, Report of Investigations 5214, 1956.

233. M. Chatterjee, J. Ray, A. Chatterjee and D. Ganguli, "Characterization of basic zirconium sulphate, a precursor for zirconia," *J. Mater. Sci. Lett.*, **8**, 548–50 (1989).

234. M. Chatterjee, J. Ray, A. Chatterjee and D. Ganguli, "High purity zirconia powders via wet chemical processing : a comparative study," *Ceram. Int.*, **18**, 337–42 (1992).

235. T. Uetsuki, K. Tanaka, M. Adachi and Y. Nakazawa, "A method for the production of zirconia from zircon," *Taikabutsu Overseas*, **6** (2), 3–9 (1986).

236. U. Syamaprasad, S. Bhattacharjee, R. K. Galgali and B. C. Mohanty, "Zirconia ceramics by thermal plasma," *Trans. Ind. Ceram. Soc.*, **52**, 35–46 (1993).

237. T. Itoh, "Particle and crystallite sizes of ZrO_2 powder obtained by the calcination of hydrous zirconia," *J. Mater. Sci. Lett.*, **4**, 431–33 (1985).

238. P. Duran, P. Recio, J. R. Jurado, C. Pascual and C. Moure, "Preparation, sintering and properties of translucent Er_2O_3-doped tetragonal zirconia," *J. Am. Ceram. Soc.*, **72**, 2088–93 (1989).

239. F.-C. Wu and S.-C. Yu, "Effects of H_2SO_4 on the crystallization and phase transformation of zirconia powder in the precipitation processes," *J. Mater. Sci.*, **25**, 970–76 (1990).

240. F.-C. Wu and S.-C. Yu, "The crystallization and phase transformation of stabilizer-doped zirconia powders with H_2SO_4 additive," *J. Cryst. Growth*, **96**, 96–100 (1989).

241. H. Hofmann, B. Michel and L. J. Gauckler, "Zirconia powder of TZP-ceramics Ti-Y-TZP"; pp. 119–29 in *Zirconia' 88, Advances in Zirconia Science and Technology*. Edited by S. Meriani and C. Palmonari. Elsevier Applied Science, London, 1989.

242. M. A. C. G. van de Graaf and A. J. Burggraaf, "Wet-chemical preparation of zirconia powders : their microstructure and behaviour"; pp. 744–65 in *Adv. Ceram.*, **12**, *Science and Technology of Zirconia II*. Edited by N. Claussen, M. Rühle and A. H. Heuer. American Ceramic Society, Columbus, OH, 1984.

243. M. A. Blesa, A. J. G. Maroto, S. I. Passaggio, N. E. Figliolia and G. Rigotti, "Hydrous zirconium dioxide : interfacial properties, the formation of monodisperse spherical particles, and its crystallization at high temperatures," *J. Mater. Sci.*, **20**, 4601–4609 (1985).

244. B. Djuricic, D. McGarry and S. Pickering, "The preparation of ultrafine ceria-stabilized zirconia particles coated with urea," *J. Mater. Sci. Lett.*, **12**, 1320–23 (1993).

245. E. Tani, M. Yoshimura and S. Somiya, "Formation of ultrafine tetragonal ZrO_2 powder under hydrothermal conditions," *J. Am. Ceram. Soc.*, **66**, 11–14 (1983).

246. L. Michalowsky and H.- D. Schnabel, "Sintering of coprecipitated MgO-partially stabilized zirconia," *Ceram. Int.*, **16**, 33–37 (1990).

247. K. Haberko, W. Pyda, M. Kuras and M. Bucko, "Tetragonal zirconia polycrystals in the MgO-Y_2O_3-ZrO_2 system (MgY-TZP)"; pp. 107–118 in *Zirconia' 88, Advances in Zirconia Science and Technology.* Edited by S. Meriani and C. Palmonari. Elsevier Applied Science, London, 1989.

248. A. Bleier and R. M. Cannon, "Nucleation and growth of uniform m-ZrO_2"; pp. 71–78 in *Mater. Res. Soc. Symp. Proc.*, **73**, *Better Ceramics Through Chemistry II.* Edited by C. J. Brinker, D. E. Clark and D. R. Ulrich. Materials Research Society, Pittsburgh, PA, 1986.

249. B. Mottet, M. Pichavant, J.-M. Beny and J.-A. Alary, "Morphology of zirconia synthesized hydrothermally from zirconium oxychloride," *J. Am. Ceram. Soc.*, **75**, 2515–19 (1992).

250. S. Komarneni, R. Roy, E. Breval, M. Ollinen and Y. Suwa, "Hydrothermal route to ultrafine powders utilizing single and diphasic gels," *Adv. Ceram. Mater.*, **1**, 87–92 (1986).

251. C. Li, I. Yamai, Y. Murase and E. Kato, "Formation of acicular monoclinic zirconia particles under hydrothermal conditions," *J. Am. Ceram. Soc.*, **72**, 1479–82 (1989).

252. E. Kato, M. Hirano and A. Nagai, "Growth of monoclinic ZrO_2 thin, flaky crystals by hydrothermal decomposition of zirconium oxide sulphate crystals," *J. Am. Ceram. Soc.*, **78**, 2259–62 (1995).

253. B. Dubois, D. Ruffier and P. Odier, "Preparation of fine, spherical yttria-stabilized zirconia by the spray pyrolysis method," *J. Am. Ceram. Soc.*, **72**, 713–15 (1989).

254. T. Yoshioka, K. Dosaka, T. Sato, A. Okuwaki, S. Tanno and T. Miura, "Preparation of spherical ceria-doped tetragonal zirconia by the spray-pyrolysis method," *J. Mater. Sci. Lett.*, **11**, 51–55 (1992).

255. D. Xiaming, L. Qingfeng and T. Yuying, "Study of phase formation in spray pyrolysis of ZrO_2 and ZrO_2-Y_2O_3 powders," *J. Am. Ceram. Soc.*, **76**, 760–62 (1993).

256. M. Chatterjee and D. Ganguli, "Preparation of pure and stabilizer (MgO, CeO_2)-added ZrO_2 powders and their crystallization behaviour," *Trans. Ind. Ceram. Soc.*, **45**, 147–51 (1986).

257. L. Rakotoson and M. Paulus, "Sintering of a freeze-dried 10 mol% Y_2O_3-stabilized zirconia"; pp. 727–32 in *Adv. Ceram.*, **12**, *Science and Technology of Zirconia II.* Edited by N. Claussen, M. Rühle and A. H. Heuer. American Ceramic Society, Columbus, OH, 1984.

258. K. S. Mazdiyasni, C. T. Lynch and J. S. Smith II, "Cubic phase stabilization of translucent yttria-zirconia at very low temperatures," *J. Am. Ceram. Soc.*, **50**, 532–37 (1967).

259. R. Guinebretiere, A. Dauger, A. Lecomte and H. Vesteghem, "Tetragonal zirconia powders from the zirconium n-propoxide–acetylacetone–water–isopropanol system," *J. Non-Cryst. Solids*, **147 & 148**, 542–47 (1992).

260. B. E. Yoldas, "Zirconium oxides formed by hydrolytic condensation of alkoxides and parameters that affect their morphology," *J. Mater. Sci.*, **21**, 1080–86 (1986).

261. B. Fegley, Jr., P. White and H. K. Bowen, "Processing and characterization of ZrO_2 and Y-doped ZrO_2 powders," *Am. Ceram. Soc. Bull.*, **64**, 1115–20 (1985).

262. G. Rinn and H. Schmidt, "Preparation of monodispersed zirconia powders from solution"; pp. 23–30 in *Ceram. Trans.*, **1**, *Ceramic Powder Science II.* Edited by G. L. Messing, E. R. Fuller, Jr. and H. Hausner, American Ceramic Society, Westerville, OH, 1988.

263. G. F. Tu, Z. T. Sui, Q. Huang and C. Z. Wang, "Sol-gel processed Y-PSZ ceramics with 5 wt% Al_2O_3," *J. Am. Ceram. Soc.*, **75**, 1032–34 (1992).

264. T. Okubo and H. Nagamoto, "Low-temperature preparation of nanostructured zirconia and YSZ by sol-gel processing," *J. Mater. Sci.*, **30**, 749–57 (1995).

265. S. Meriani and G. Soraru, "Ultrafine ceria zirconia powders obtained via metallorganic precursors"; pp. 547-54 in. *Ceramic Powders.* Edited by P. Vincenzini. Elsevier, Amsterdam, 1983.

266. M. Chatterjee, J. Ray, A. Chatterjee, D. Ganguli, S. V. Joshi and M. P. Srivastava, "Thermal barrier coatings from sol-gel-derived spray-grade Y_2O_3-ZrO_2 microspheres," *J. Mater. Sci.*, **28**, 2803–2807 (1993).

267. F. G. Sherif and L.-J. Shyu, "Emulsion precipitation of yttria-stabilized zirconia for plasma spray coatings," *J. Am. Ceram. Soc.*, **74**, 375–80 (1991).

268. S. D. Ramamurthi, Z. Xu and D. A. Payne, "Nanometer-sized ZrO_2 particles prepared by a sol-emulsion-gel method," *J. Am. Ceram. Soc.*, **73**, 2760–63 (1990).

269. A. Srivastava and M. K. Dongare, "Low-temperature preparation of tetragonal zirconia," *Mater. Lett.*, **5**, 111-15 (1987).

270. M. Yashima, K. Ohtake, M. Kakihana and M. Yoshimura, "Synthesis of metastable tetragonal (t') zirconia-ceria solid solutions by the polymerized complex method," *J. Am. Ceram. Soc.*, **77**, 2773–76 (1994).

271. V. G. Chukhlantsev and Y. M. Galkin, "Thermal decomposition of basic zirconium sulphate," *Russ. J. Inorg. Chem.*, **18**, 770–71 (1973).

272. J. L. Shi and J. H. Gao, "Preparation of spherical zirconium salt particles by homogeneous precipitation," *J. Mater. Sci.*, **30**, 793–99 (1995).

273. K. S. Mazdiyasni, C. T. Lynch and J. S. Smith, "Preparation of ultra-high-purity submicron refractory oxides," *J. Am. Ceram. Soc.*, **48**, 372–75 (1965).

274. N. Arul Dhas and K. C. Patil, "Preparation of magnesia-stabilized zirconia powders prepared by a combustion route," *J. Mater. Sci. Lett.*, **12**, 1844–47 (1993).

275. R. Ravelle-Chapuis, M. G. Blanchin, M. Jebrouni and B. Durand, "Characterization of pure ZrO_2 powders processed from reactions in molten salt mixtures"; pp. 2.97–2.101 in *Euro-Ceramics*, **2**. Edited by G. de With, R. A. Terpstra and R. Metselaar. Elsevier Applied Science, London, 1989.

276. M. Descemond, C. Brodhag, F. Thevenot, B. Durand, M. Jebrouni and M. Roubin, "Characteristics and sintering behaviour of 3 mol % Y_2O_3-ZrO_2 powders synthesized by reaction in molten salts," *J. Mater. Sci.*, **28**, 2283–88 (1993).

277. J. Whitehead, "Titanium compounds (inorganic)"; pp. 131–76 in *Kirk-Othmer Encyclopedia of Chemical Technology*, Vol. 23 (3rd ed.). John Wiley, New York, 1983.

278. Anon. *Ceram. Ind.*, **144** (1), 145–46 (1995).

279. Y. Suyama and A. Kato, "TiO_2 produced by vapour-phase oxygenolysis of $TiCl_4$," *J. Am. Ceram. Soc.*, **59**, 146–49 (1976).

280. M. K. Akhtar, Y. Xiong and S. E. Pratsinis, "Vapour synthesis of titania powder by titanium tetrachloride oxidation," *A I Ch E J.*, **37**, 1567–70 (1991).

281. T. Oyama, Y. Iimura and K. Takeuchi, "Synthesis of rutile and anatase TiO_2 fine particles by laser-ignited vapour-phase reaction," *J. Mater. Sci. Lett.*, **15**, 594–96 (1996).

282. S. Vemury and S. E. Pratsinis, "Dopants in flame synthesis of titania," *J. Am. Ceram. Soc.*, **78**, 2984–92 (1995).

283. M. Vallet-Regi, J. Pena, A. Martinez and J. M. Gonzalez-Calbet, "Selection of structural type and particle size in titanium (IV) oxide," *J. Mater. Res.*, **8**, 2336–43 (1993).

284. W. Bauer and G. Tomandl, "Preparation of spherical TiO_2 particles by an emulsion method using $TiCl_4$," *Ceram. Int.*, **20**, 189–93 (1994).

285. V. Chhabra, V. Pillai, B. K. Mishra, A. Morrone and D. O. Shah, "Synthesis, characterization and properties of microemulsion mediated nanophase TiO_2 particles," *Langmuir*, **11**, 3307–3311 (1995).

286. M. Yoshimura, H. Ohira and S. Somiya, "Formation of TiO_2 and ZnO powders by hydrothermal oxidation of Ti and Zn metals," *Rept. Res. Lab. Eng. Mater., Tokyo Inst. Technol.*, **12**, 59–72 (1987).

287. Y.-T. Qian, Q.-W. Chen and Z.-W. Chen, "Preparation of ultrafine powders of TiO_2 by hydrothermal H_2O_2 oxidation starting from metallic Ti," *J. Mater. Chem.*, **3**, 203–205 (1993).

288. T. Ogihara, T. Yanagawa, N. Ogata and K. Yoshida, "Formation of monodispersed oxide particles by hydrolysis of metal alkoxide in octanol/acetonitrile solutions," *J. Ceram. Soc. Japan*, **101**, 315–320 (1993). [in Japanese]

289. J. H. Jean, D. M. Goy and T. A. Ring, "Continuous production of narrow-sized and unagglomerated TiO_2 powders," *Am. Ceram. Soc. Bull.*, **66**, 1517–20 (1987).

290. A. Van Zyl, P. M. Smit and A. I Kingdon, "A chemical reactor synthesis for TiO_2 and ZrO_2 powder synthesis," *Mater. Sci. Eng.*, **78**, 217–22 (1986).

291. T. Ogihara, M. Ikeda, M. Kato and N. Mizutani, "Continuous processing of monodispersed titania powders," *J. Am. Ceram. Soc.*, **72**, 1598–1601 (1989).

292. E. Haro-Poniatowski, R. Rodriguez-Talavera, M. de la Cruz Heredia, O. Cano-Corona and R. Arroyo-Murillo, "Crystallization of nanosized titania particles prepared by the sol-gel process," *J. Mater. Res.*, **9**, 2102–2108 (1994).

293. F. Kirkbir and H. Komiyama, "Continuous production of fine TiO_2 powders by vapour-phase hydrolysis of titanium tetraisopropoxide," *Adv. Ceram. Mater.*, **3**, 511–15 (1988).

294. H. Kominami, Y. Takada, H. Yamagiwa, Y. Kera, M. Inoue and T. Inui, "Synthesis of thermally stable nanocrystalline anatase by high-temperature hydrolysis of titanium alkoxide with water dissolved in organic solvent from gas phase," *J. Mater. Sci. Lett.*, **15**, 197–200 (1996).

295. A. Kumar and R. Roy, "RESA – a wholly new process for fine oxide powder preparation," *J. Mater. Res.*, **3**, 1373–77 (1988).

296. A. N. Copp, "Magnesia/magnesite," *Am. Ceram. Soc. Bull.*, **72** (6), 107–109 (1993).

297. R. J. Hall and D. R. F. Spencer, "A review of the production and properties of sea-water magnesia," *Interceram*, **22**, 212–18 (1973).

298. R. A. Brown, "Sintering in very pure magnesium oxide and magnesium oxide containing vanadium," *Am. Ceram. Soc. Bull.*, **44**, 483–87 (1965).

299. I. F. Guilliatt and N. H. Brett, "The effect of impurities on the powder properties of $Mg(OH)_2$ decomposed in vacuo," *Trans. Brit. Ceram. Soc.*, **69**, 1–7 (1970).

300. T. J. Gardner and G. L. Messing, "Preparation of MgO powder by evaporative decomposition of solutions," *Am. Ceram. Soc. Bull.*, **63**, 1498–1501 (1984).

301. C. A. Handwerker, R. M. Cannon and R. L. Coble, "Final-stage sintering of MgO"; pp. 619–43 in *Adv. Ceram.*, **10**, *Structure and Properties of MgO and Al_2O_3 Ceramics*. Edited by W. D. Kingery. American Ceramic Society, Columbus, OH, 1984.

302. R. Portillo, T. Lopez, R. Gomez, Bokhimi, A. Morales and O. Novaro, "Magnesia synthesis via sol-gel : structure and reactivity," *Langmuir*, **12**, 40–44 (1996).

303. A. Nishida, A. Ueki and K. Yoshida, "Process effects on the particle size of magnesium oxide powders synthesized by vapour-phase oxidation," pp. 265–69 in *Adv. Ceram.*, **21**, *Ceramic Powder Science*. Edited by G. L. Messing, K. S. Mazdiyasni, J. W. McCauley and R. A. Haber. American Ceramic Society, Westerville, OH, 1987.

304. T. Watari, K. Nakayoshi and A. Kato, "Preparation of submicron magnesium oxide powders by vapour-phase reaction of magnesium vapour and oxygen," *Nippon Kagaku Kaishi*, 1075–76 (1984). [in Japanese]

305. K. Yoshizawa, Y. Sugoh and Y. Ochi, "Controlled growth of monodispersed spherical silica by a new synthetic process"; pp. 125–31 in *Sci. Ceram.*, **14**. Edited by D. Taylor. The Institute of Ceramics, Stoke-on-Trent, U. K., 1988.

306. H. Giesche, "Synthesis of monodispersed silica powders I. Particle properties and reaction kinetics," *J. Eur. Ceram. Soc.*, **14**, 189–204 (1994); also "II. Controlled growth reaction and continuous production process," *J. Eur. Ceram. Soc.*, **14**, 205–214 (1994).

307. G. H. Bogush and C. F. Zukoski IV, "Studies on the formation of monodisperse silica powders"; pp. 477–86 in *Ultrastructure Processing of Advanced Ceramics*. Edited by J. D. Mackenzie and D. R. Ulrich, John Wiley, New York, 1988.

308. T. Kawaguchi and K. Ono, "Spherical silica gels precipitated from acid catalyzed TEOS solutions," *J. Non-Cryst. Solids*, **121**, 383–88 (1990).

309. H. Yamashita, M. Demiya, H. Mori and T. Maekawa, "Synthesis of microporous silica-gel particles in W/O emulsion and an application to high-performance liquid chromatography," *J. Ceram. Soc. Japan*, **100**, 1444–47 (1992). [in Japanese]

310. M. A. Butler, P. F. James and J. D. Jackson, "An emulsion method for producing fine, low density, high surface area silica powder from alkoxides," *J. Mater. Sci.*, **31**, 1675–80 (1996).

311. B. Karmakar, G. De, D. Kundu and D. Ganguli, "Silica microspheres from the system tetraethyl orthosilicate-acetic acid-water," *J. Non-Cryst. Solids*, **135**, 29–36 (1991).

312. J. Y. Ding and D. E. Day, "Preparation of silica glass microspheres by sol-gel processing," *J. Mater. Res.*, **6**, 168–74 (1991).

313. *Cab-O-Sil Fumed Silica, Properties and Functions.* Product booklet of Cabot Corporation, U. S. A. (1987).

314. P. C. Aitcin, P. Pinsonneault and D. M. Roy, "Physical and chemical characterization of condensed silica fumes," *Am. Ceram. Soc. Bull.*, **63**, 1487–91 (1984).

315. M. Hirano and E. Kato, "Hydrothermal synthesis of cerium (IV) oxide," *J. Am. Ceram. Soc.*, **79**, 777–80 (1996).

316. M. D. Rasmussen, G. W. Jordan, M. Akinc, O. Hunter, Jr. and M. F. Berard, "Influence of precipitation procedure on sinterability of Y_2O_3 prepared from hydroxide precursor," *Ceram. Int.*, **9**, 59–60 (1983).

317. B. Djuricic, D. Kolar and M. Memic, "Synthesis and properties of Y_2O_3 powder obtained by different methods," *J. Eur. Ceram. Soc.*, **9**, 75–82 (1992).

318. M. Akinc and A. Celikkaya, "Preparation of yttria powders by emulsion precipitation"; pp. 57–67 in *Adv. Ceram.*, **21**, *Ceramic Powder Science.* Edited by G. L. Messing, K. S. Mazdiyasni, J. W. McCauley and R. A. Haber. American Ceramic Society, Westerville, OH, 1987.

319. T. Hours, P. Bergez, J. Charpin, A. Larbot, C. Guizard and L. Cot, "Preparation and characterization of yttrium oxide by a sol-gel process," *Am. Ceram. Soc. Bull.*, **71**, 200–203 (1992).

320. A. J. Rulison and R. C. Flagan, "Synthesis of yttria powders by electrospray pyrolysis," *J. Am. Ceram. Soc.*, **77**, 3244–50 (1994).

321. A. L. Micheli, "Synthesis of reactive yttria powders using an aqueous organic polymer precursor"; pp. 102–109 in *Ceram. Trans.*, **1**, *Ceramic Powder Science II.* Edited by G. L. Messing, E. R. Fuller, Jr. and H. Hausner. American Ceramic Society, Westerville, OH, 1988.

322. Y. Minagawa and F. Yajima, "Preparative method of fine powder of yttrium (III) oxide by thermal decomposition of $NH_4Y(C_2O_4)_2$. H_2O fine crystal by reaction of acidic solution of yttrium oxalate and aqueous ammonia," *Bull. Chem. Soc. Japan*, **63**, 378–82 (1990).

323. S. Ekambaram and K. C. Patil, "Combustion synthesis of yttria," *J. Mater. Chem.*, **5**, 905–908 (1995).

324. E. Matijevic, "Preparation and interactions of colloids of interest in ceramics"; pp. 429–42 in *Ultrastructure Processing of Advanced Ceramics.* Edited by J. D. Mackenzie and D. R. Ulrich. John Wiley, New York, 1988.

325. E. Matijevic and W. P. Hsu, "Preparation and properties of monodispersed colloidal particles of lanthanide compounds : I. Gadolinium, europium, terbium, samarium and cerium (III)", *J. Colloid Interface Sci.*, **118**, 506–523 (1987).

326. P.-L. Chen and I.-W. Chen, "Reactive cerium (IV) oxide powders by the homogeneous precipitation method," *J. Am. Ceram. Soc.*, **76**, 1577–83 (1993).

327. M. N. Rahaman and Y. C. Zhou, "Effect of solid solution additives on the sintering of ultra-fine CeO_2 powders," *J. Eur. Ceram. Soc.*, **15**, 939–50 (1995).

328. E. Verdon, M. Devalette and G. Demazeau, "Solvothermal synthesis of cerium dioxide microcrystallites : effect of the solvent," *Mater. Lett.*, **25**, 127–31 (1995).

329. E. Tani, M. Yoshimura and S. Somiya, "Crystallization and crystal growth of CeO_2 under hydrothermal conditions," *J. Mater. Sci.Lett.*, **1**, 461–62 (1982).

330. A. S. Perl, "Zinc oxide," *Am. Ceram. Soc. Bull.*, **72** (6), 122–23 (1993).

331. Anon, "Zinc oxide," *Ceram. Ind.*, **144** (1), 149–50 (1995).

332. E. Sonder, T. C. Quinby and D. L. Kinser, "ZnO varistors made from powders produced using a urea process," *Am. Ceram. Soc. Bull.*, **65**, 665–68 (1986).

333. S. Hishita, Y. Yao and S.-I. Shirasaki, "Zinc oxide varistors made from powders prepared by amine processing," *J. Am. Ceram. Soc.*, **72**, 338–40 (1989).

334. S. M. Haile, D. W. Johnson, Jr., G. H. Wiseman and H. K. Bowen, "Aqueous precipitation of spherical zinc oxide powders for varistor applications," *J. Am. Ceram. Soc.*, **72**, 2004–2008 (1989).

335. Q. Zhong and E. Matijevic, "Preparation of uniform zinc oxide colloids by controlled double-jet precipitation," *J. Mater. Chem.*, **6**, 443–47 (1996).

336. H. Nishizawa, T. Tani and K. Matsuoka, "Crystal growth of ZnO by hydrothermal decomposition of Zn-EDTA," *J. Am. Ceram. Soc.*, **67**, C-98–C-100 (1984).

337. D. Jezequel, J. Guenot, N. Jouini and F. Fievet, "Preparation and morphological characterization of fine, spherical, monodisperse particles of ZnO"; pp. 339–42 in *Mater. Sci. Forum*, **152–153**, *Soft Chemistry Routes to New Materials*. Edited by J. Rouxel, M. Tournoux and R. Brec. Trans Tech Publications, Switzerland, 1994.

338. E. M. Vogel, N. C. Andreadakis, W. E. Quinn and T. J. Nelson, "Materials for display devices based on metallo-organic precursors"; pp. 131–37 in *Adv. Ceram.*, **21**, *Ceramic Powder Science*. Edited by G. L. Messing, K. S. Mazdiyasni, J. W. McCauley and R. A. Haber. American Ceramic Society, Westerville, OH, 1987.

339. R. H. Heistand II and Y.-H. Chia, "Synthesis of submicron, narrow size distribution spherical zincite"; pp. 93–98 in *Mater. Res. Soc. Symp. Proc.*, **73**, *Better Ceramics Through Chemistry II*. Edited by C. J. Brinker, D. E. Clark and D. R. Ulrich. Materials Research Society, Pittsburgh, PA, 1986.

340. K. Kamata, H. Hosono, Y. Maeda and K. Miyokawa, "Synthesis of zinc oxide powder by hydrolysis of bis(acetylacetonato)–zinc (II) in aqueous solution," *Chem. Lett.*, 2021–2022 (1984).

341. Y. Suyama, S. Takemiya and A. Kato, "Synthesis of ultrafine ZnO powders by vapour phase oxidation of Zn vapour," *Yogyo-Kyokai-Shi*, **93**, 105–107 (1985). [in Japanese]

342. Y. Suyama, Y. Tomokiyo, T. Manabe and E. Tanaka, "Shape and structure of zinc oxide particles prepared by vapour-phase oxidation of zinc vapour," *J. Am. Ceram. Soc.*, **71**, 391–95 (1988).

343. W. E. Cameron, "Composition and cell dimensions of mullite," *Am. Ceram. Soc. Bull.*, **56**, 1003–1007, 1011 (1977).

344. R. X. Fischer, H. Schneider and D. Voll, "Formation of aluminium rich 9 : 1 mullite and its transformation to low alumina mullite upon heating," *J. Eur. Ceram. Soc.*, **16**, 109–113 (1996).

345. J. A. Pask and A. P. Tomsia, "Formation of mullite from sol-gel mixtures and kaolinite," *J. Am. Ceram. Soc.*, **74**, 2367–73 (1991).

346. P. Boch, T. Chartier and P. D. D. Rodrigo, "High-purity mullite ceramics by reaction-sintering"; pp. 353–74 in *Ceram. Trans.*, **6**, *Mullite and Mullite Matrix Composites*. Edited by S. Somiya, R. F. Davis and J. A. Pask. American Ceramic Society, Westerville, OH, 1990.

347. B. B. Ghate, D. P. H. Hasselman and R. M. Spriggs, "Synthesis and characterization of high purity, fine grained mullite," *Am. Ceram. Soc. Bull.*, **52**, 670–72 (1973).

348. M. D. Sacks and J. A. Pask, "Sintering of mullite-containing materials : II, effect of agglomeration," *J. Am. Ceram. Soc.*, **65**, 70-77 (1982).

349. M. G. M. U. Ismail, Z. Nakai and S. Somiya, "Microstructure and mechanical properties of mullite prepared by the sol-gel method," *J. Am. Ceram. Soc.*, **70**, C-7–C-8 (1987).

350. M. Mizuno and H. Saito, "Preparation of highly pure mullite powder," *J. Am. Ceram. Soc.*, **72**, 377–82 (1989).

351. J. G. Liu and D. L. Wilcox, Sr., "Factors influencing the formation of hollow ceramic microspheres by water extraction of colloidal droplets," *J. Mater. Res.*, **10**, 84–94 (1995).

352. I. Jaymes and A. Douy, "Homogeneous precipitation of mullite precursors," *J. Sol-Gel Sci. Tech.*, **4**, 7–13 (1995).

353. B. L. Metcalfe and J. H. Sant, "The synthesis, microstructure and physical properties of high purity mullite," *Trans. J. Brit. Ceram. Soc.*, **74**, 193–201 (1975).

354. S. Kanzaki, H. Tabata, T. Kumazawa and S. Ohta, "Sintering and mechanical properties of stoichiometric mullite," *J. Am. Ceram. Soc.*, **68**, C-6–C-7 (1985).

355. K. A. Moore, J. Cesarano III, D. M. Smith and T. T. Kodas, "Synthesis of submicrometer mullite powder via high-temperature aerosol decomposition," *J. Am. Ceram. Soc.*, **75**, 213–15 (1992).

356. K. S. Mazdiyasni and L. M. Brown, "Synthesis and mechanical properties of stoichiometric aluminium silicate (mullite)," *J. Am. Ceram. Soc.*, **55**, 548–52 (1972).

357. S. Mitachi, M. Matsuzawa, K. Kaneko, S. Kanzaki and H. Tabata, "Characterization of SiO_2-Al_2O_3 powders prepared from metal alkoxides"; pp. 275–86 in *Ceram. Trans.*, **6**, *Mullite and Mullite Matrix Composites*. Edited by S. Somiya, R. F. Davis and J. A. Pask. American Ceramic Society, Westerville, OH, 1990.

358. H. Suzuki, H. Saito, Y. Tomokiyo and Y. Suyama, "Processing of ultrafine mullite powder through alkoxide route"; pp. 263–74 in *Ceram. Trans.*, **6**, *Mullite and Mullite Matrix Composites*. Edited by S. Somiya, R. F. Davis and J. A. Pask. American Ceramic Society, Westerville, OH, 1990.

359. D. Y. Jeng and M. N. Rahaman, "Sintering and crystallization of mullite powder prepared by sol-gel processing," *J. Mater. Sci.*, **28**, 4904–4909 (1993).

360. H. Schneider, L. Merwin and A. Sebald, "Mullite formation from non-crystalline precursors," *J. Mater. Sci.*, **27**, 805–812 (1992).

361. H. Schneider, B. Saruhan, D. Voll, L. Merwin and A. Sebald, "Mullite precursor phases," *J. Eur. Ceram. Soc.*, **11**, 87–94 (1993).

362. L. A. Paulick, Y.-F. Yu and T.-I. Mah, "Ceramic powders from metal alkoxide precursors"; pp. 121–29 in *Adv. Ceram.*, **21**, *Ceramic Powder Science.* Edited by G. L. Messing, K. S. Mazdiyasni, J. W. McCauley and R. A. Haber. American Ceramic Society, Westerville, OH, 1987.

363. S. Somiya, M. Yoshimura, M. Suzuki and T. Yamaguchi, "Mullite powder from hydrothermal processing"; pp. 287–310 in *Ceram. Trans.*, **6**, Mullite and Mullite Martix Composites. Edited by S. Somiya, R. F. Davis and J. A. Pask. American Ceramic Society, Westerville, OH, 1990.

364. F. J. Berry, M. W. Booth, P. M. Marquis, M. Mortimer, M. R. Piramon and C. B. Ponton, "Formation of mullite from hydrothermally processed sols : an X-ray powder diffraction and magic angle spinning NMR investigation," *J. Mater. Chem.*, **3**, 965–68 (1993).

365. G.-Y. Meng and R. A. Huggins, "A new chemical method for preparation of both pure and doped mullite," *Mater. Res. Bull.*, **18**, 581–88 (1983).

366. A. Douy, "Organic gels in the preparation of silicate powders : examples of mullite and cordierite"; pp. 585–94 in *Chemical Processing of Advanced Materials.* Edited by L. L. Hench and J. K. West. John Wiley, New York, 1992.

367. C. S. Hong, P. Ravindranathan, D. K. Agrawal and R. Roy, "Synthesis and sintering of mullite powders by the decomposition/combustion of aluminium nitrate-amorphous fumed silica-urea mixtures," *J. Mater. Sci. Lett.*, **13**, 1072–1075 (1994).

368. P. E. D. Morgan, "Structuring chemical technology to produce cost-effective ceramic products on the large scale," *Am. Ceram. Soc. Bull.*, **72** (7), 65–70 (1993).

369. S. Hori and R. Kurita, "Characterization and sintering of Al_2O_3-SiO_2 powders formed by chemical vapour deposition"; pp. 311–22 in *Ceram. Trans.*, **6**, *Mullite and Mullite Martix Composites.* Edited by S. Somiya, R. F. Davis and J. A. Pask. American Ceramic Society, Westerville, OH, 1990.

370. W. T. Bakker and J. G. Lindsay, "Reactive magnesia spinel, preparation and properties," *Am. Ceram. Soc. Bull.*, **46**, 1094–1097 (1967).

371. J. Beretka and T. Brown, "Effect of particle size on the kinetics of the reaction between magnesium and aluminium oxides," *J. Am. Ceram. Soc.*, **66**, 383–88 (1983).

372. R. J. Bratton, "Coprecipitates yielding $MgAl_2O_4$ spinel powders," *Am. Ceram. Soc. Bull.*, **48**, 759–62 (1969).

373. R. J. Bratton, "Characterization and sintering of reactive $MgAl_2O_4$ spinel," *Am. Ceram. Soc. Bull.*, **48**, 1069–1075 (1969).

374. G. Gusmano, P. Nunziante, E. Traversa and G. Chiozzini, "The mechanism of $MgAl_2O_4$ spinel formation from the thermal decomposition of coprecipitated hydroxides," *J. Eur. Ceram. Soc.*, **7**, 31–39 (1991).

375. H. Bhattacharya and B. N. Samaddar, "Formation of a nonstoichiometric spinel on heating hydrous magnesium aluminate," *J. Am. Ceram. Soc.*, **61**, 279–80 (1978).

376. G. G. Galimov, N. F. Seliverstov, V. A. Ryabin, T. N. Zhitkova, N. A. Berg, L. N. Kalitina, I. A. Leont'eva and Yu. I. Savchenko, "Preparation of magnesium aluminate from sodium aluminate solutions and magnesium salts," *Inorg. Mater.*, **22**, 559–63 (1986).

377. Z Nakagawa, K. Hamano, M. Sakaguchi and S. Kanzaki, "Characterization and sinterability of Mg-Al spinel powders prepared with a thermal decomposition of a freeze-dried sulphate," *Yogyo-Kyokai-Shi*, **90**, 313–19 (1982). [in Japanese]

378. P. W. D. Mitchell, "Chemical method of preparing $MgAl_2O_4$ spinel," *J. Am. Ceram. Soc.*, **55**, 484 (1972).

379. M. Sugiura and O. Kamigaito, "Characterization and formation process of spinel ($MgAl_2O_4$) prepared by alkoxide ($MgAl_2(i\text{-}OC_3H_7)_8$) method," *Yogyo-Kyokai-Shi*, **92**, 605–611 (1984).

380. J.-F. Pasquier, S. Komarneni and R. Roy, "Synthesis of $MgAl_2O_4$ spinel : seeding effects on formation temperature," *J. Mater. Sci.*, **26**, 3797–3802 (1991).

381. D. R. Messier and G. E. Gazza, "Synthesis of $MgAl_2O_4$ and $Y_3\,Al_5O_{12}$ by thermal decomposition of hydrated nitrate mixtures," *Am. Ceram. Soc. Bull.*, **51**, 692–94, 97 (1972).

382. J. Nowotny and M. Rekas, "Dielectric ceramic materials based on alkaline earth metal titanates"; pp. 45–143 in *Key Eng. Mater.*, **66 & 67**. *Electronic Ceramic Materials.* Edited by J. Nowotny. Trans Tech Publications, Zurich, 1992.

383. A. D. Hilton and R. Frost, "Recent developments in the manufacture of barium titanate powders"; pp. 145–83 in *Key Eng. Mater.*, **66 & 67**. *Electronic Ceramic Materials.* Edited by J. Nowotny. Trans Tech Publications, Zurich, 1992.

384. P. P. Phule and S. H. Risbud, "Low-temperature synthesis and processing of electronic materials in the $BaO\text{-}TiO_2$ system," *J. Mater. Sci.*, **25**, 1169–83 (1990).

385. L. K. Templeton and J. A. Pask, "Formation of $BaTiO_3$ from $BaCO_3$ and TiO_2 in air and in CO_2," *J. Am. Ceram. Soc.*, **42**, 212–16 (1959).

386. A. Beauger, J. C. Mutin and J. C. Niepce, "Synthesis reaction of metatitanate $BaTiO_3$. Part 1. Effect of the gaseous atmosphere upon the thermal evolution of the system $BaCO_3$–TiO_2," *J. Mater Sci.*, **18**, 3041–3046 (1983); also "Part 2. Study of solid-solid reaction interfaces," *J. Mater. Sci.*, **18**, 3543–50 (1983).

387. M. S. H. Chu and A. W. I. M. Rae, "Manufacturing dielectric powders," *Am. Ceram. Soc. Bull.*, **74** (6), 69–72 (1995).

388. D. Hennings, "Review of chemical preparation routes for barium titanate"; pp. 1–10 in *Brit. Ceram. Proc.*, **41**, *Electroceramics.* Edited by A. J. Moulson, J. Binner and R. Morrell. The Institute of Ceramics, Stoke-on-Trent, U. K., 1989.

389. D. K. Gallagher and J. Thomson, Jr., "Thermal analysis of some barium and strontium titanyl oxalates," *J. Am. Ceram. Soc.*, **48**, 644–47 (1965).

390. K. Kudaka, K. Iizumi and K. Sasaki, "Preparation of barium titanyl oxalate tetrahydrate," *Am. Ceram. Soc. Bull.*, **61**, 1236 (1982).

391. H. Yamamura, A. Watanabe, S. Shirasaki, Y. Moriyoshi and M. Tanada, "Preparation of barium titanate by oxalate method in ethanol solution," *Ceram. Int.*, **11**, 17–22 (1985).

392. V. W. Day, T. A. Eberspacher, M. H. Frey, W. G. Klemperer, S. Liang and D. A. Payne, "Barium titanium glycolate : a new barium titanate powder precursor," *Chem. Mater.*, **8**, 330–32 (1996).

393. S. Wada, T. Suzuki and T. Noma, "Preparation of barium titanate fine particles by hydrothermal method and their characterization," *J. Ceram. Soc. Japan*, **103**, 1220–27 (1995).

394. C.-T. Xia, E.-W. Shi, W.-Z. Zhong and J.-K. Guo, "Preparation of $BaTiO_3$ by the hydrothermal method," *J. Eur. Ceram. Soc.*, **15**, 1171–76 (1995).

395. U. S. Patent 4,832,939 (May, 1989).

396. D. Hennings and S. Schreinemacher, "Characterization of hydrothermal barium titanate," *J. Eur. Ceram. Soc.*, **9**, 41–46 (1992).

397. N. G. Eror and H. U. Anderson, "Polymeric precursor synthesis of ceramic materials"; pp. 571–77 in *Mater. Res. Soc. Symp. Proc.*, **73**, *Better Ceramics Through Chemistry II*. Edited by C. J. Brinker, D. E. Clark and D. R. Ulrich. Materials Research Society, Pittsburgh, PA, 1986.

398. P. A. Lessing, "Mixed-cation oxide powders via polymeric precursors," *Am. Ceram. Soc. Bull.*, **68**, 1002–1007 (1989).

399. S. Kumar and G. L. Messing, "Synthesis of barium titanate by a basic pH Pechini process"; pp. 95–100 in *Mater. Res. Soc. Symp. Proc.*, **271**, *Better Ceramics Through Chemistry V*. Edited by M. J. Hampden-Smith, W. G. Klemperer and C. J. Brinker. Materials Research Society, Pittsburgh, PA, 1992.

400. D. Hennings and W. Mayr, "Thermal decomposition of (Ba, Ti) citrates into barium titanate," *J. Solid State Chem.*, **26**, 329–38 (1978).

401. J. P. Coutures, P. Odier and C. Proust, "Barium titanate formation by organic resins formed with mixed citrate," *J. Mater. Sci.*, **27**, 1849–56 (1992).

402. C. Miot, C. Proust and E. Husson, "Dense ceramics of $BaTiO_3$ produced from powders prepared by a chemical process," *J. Eur. Ceram. Soc.*, **15**, 1163–70 (1995).

403. N. J. Ali and S. J. Milne, "Synthesis and properties of barium titanate powder derived from a catechol complex," *Brit. Ceram. Trans. J.*, **86**, 113–17 (1987).

404. J. C. Bernier, "Sol -gel processing for the synthesis of powders for dielectrics," *Powder Metallurgy Int.*, **18**, 164–68 (1986).

405. K. W. Kirby, "Alkoxide synthesis techniques for $BaTiO_3$," *Mater. Res. Bull.*, **23**, 881–90 (1988).

406. Y.-S. Her, E. Matijevic and M. C. Chon, "Preparation of well-defined colloidal barium titanate crystals by the controlled double-jet precipitation," *J. Mater. Res.*, **10**, 3106–3114 (1995).

407. P. K. Gallagher, F. Schrey and F. V. DiMarcello, "Preparation of semiconducting titanates by chemical methods," *J. Am. Ceram. Soc.*, **46**, 359–65 (1963).

408. S. Bhattacharjee, M. K. Paria and H. S. Maiti, "Occurrence of excess titania in strontium titanate prepared by the oxalate precipitation route," *Ceram. Int.*, **18**, 295–300 (1992).

409. B. A. Tuttle and J. A. Voigt, "Powder synthesis of $SrTiO_3$ boundary layer capacitor materials"; pp. 62–69 in *Ceram. Trans.*, **1**, *Ceramic Powder Science II*. Edited by G. L. Messing, E. R. Fuller, Jr. and H. Hausner. American Ceramic Society, Westerville, OH, 1988.

410. S. L. Peschke, M. Ciftcioglu, D. H. Doughty and J. A. Voigt, "Preparation of strontium titanate powders by decomposition of polymeric precursors"; pp. 101–106 in *Mater. Res. Soc. Symp. Proc.*, **271**, *Better Ceramics Through Chemistry V.* Edited by M. J. Hampden-Smith, W. G. Klemperer and C. J. Brinker. Materials Reserach Society, Pittsburgh, PA, 1992.

411. G. Pfaff, "Peroxide route to synthesize strontium titanate powders of different composition," *J. Eur. Ceram. Soc.*,. **9**, 121–25 (1992).

412. J. M. G. Amores, V. S. Escribano, M. Daturi and G. Busca, "Preparation, characterization and surface structure of coprecipitated high-area Sr_xTiO_{2+x} (0 £ x £ 1) powders," *J. Mater. Chem.*, **6**, 879–86 (1996).

413. H. K. Varma, P. K. Pillai, M. M. Sreekumar, K. G. K. Warrier and A. D. Damodaran, "Strontium titanate prepared by spray drying of redispersed metal alkoxide gel," *Brit. Ceram. Trans. J.*, **90**, 189–91 (1991).

414. M. Amala Sekar, G. Dhanaraj, H. L. Bhat and K. C. Patil, "Synthesis of fine-particle titanates by the pyrolysis of oxalate precursors," *J. Mater. Sci. Mater. Electron.*, **3**, 237–39 (1992).

415. C. M. Jimenez, G. F. Arroyo and L. Del Olmo Guillen, "Reactivity studies in the formation of lead titanate and lead zirconate from coprecipitation obtained mixtures"; pp. 565–74 in *Ceramic Powders*. Edited by P. Vincenzini. Elsevier, Amsterdam, 1983.

416. Y. S. Rao and C. S. Sunandana, "Low-temperature synthesis of lead zirconate," *J. Mater. Sci. Lett.*, **11**, 595–97 (1992).

417. S. Li, R. A. Condrate Sr., J. D. Jang and R. M. Spriggs, "FTIR and Raman spectral study of the preparation of lead zirconate ($PbZrO_3$) by a sol-gel process in a non-flowing air atmosphere," *J. Mater. Sci.*, **24**, 3873–77 (1989).

418. D. M. Ibrahim and H. W. Hennicke, "Preparation of lead zirconate by a sol-gel method Part I. Sample preparation," *Trans. J. Brit. Ceram. Soc.*, **80**, 18–22 (1981).

419. M. Deri, *Ferroelectric Ceramics*. Maclaren and Sons Ltd., London, 1966.

420. G. R. Fox, E. Breval and R. E. Newnham, "Crystallization of nanometre-size coprecipitated $PbTiO_3$ powders," *J. Mater. Sci.*, **26**, 2566–72 (1991).

421. S. Bhattacharjee, M. K. Paria and H. S. Maity, "Preparation of $PbTiO_3$ powder through oxalate precipitation route," *Mater. Lett.*, **13**, 130–34 (1992).

422. D. J. Watson, C. A. Randall, R. E. Newnham and J. H. Adair, "Hydrothermal formation diagram in the lead titanate system"; pp. 154–62 in *Ceram. Trans.*, **1**, *Ceramic Powder Science II*. Edited by G. L. Messing, E. R. Fuller, Jr. and H. Hausner. Amercian Ceramic Society, Westerville, OH, 1988.

423. D. H. H. Quon, S. S. B. Wang and T. A. Wheat, "Hydrothermal synthesis of lead titanate," *Interceram*, **41**, 257–59 (1992).

424. H. Cheng, J. Ma, Z. Zhao, D. Qiang, Y. Li and X. Yao, "Hydrothermal synthesis of acicular lead titanate fine powders," *J. Am. Ceram. Soc.*, **75**, 1123–28 (1992).

425. O. Yamaguchi, A. Narai, T. Komatsu and K. Shimizu,"Crystallization and transformation of distorted cubic $PbTiO_3$," *J. Am. Ceram. Soc.*, **69**, C-256–C-257 (1986).

426. M. L. Calzada and L. Del Olmo, "Sol-gel processing by inorganic route to obtain a TiO_2-PbO xerogel as ceramic precursor," *J. Non-Cryst Solids*, **121**, 413–16 (1990).

427. T. W. Dekleva, J. M. Hayes, L. E. Cross and G. L. Geoffroy, "Sol-gel processing of lead titanate in 2-methoxyethanol : Investigations into the nature of the prehydrolyzed solutions," *J. Am. Ceram. Soc.*, **71** , C-280–C-282 (1988).

428. V. M. McNamara, "A wet chemical method for the preparation of oxide mixtures applicable to electronic ceramics," *J. Canad. Ceram. Soc.*, **34**, 103–120 (1965).

429. Y. Yoshikawa and K. Tsuzuki, "Crystallization of fine, chemically prepared lead lanthanum zirconate titanate powders at low temperature," *J. Am. Ceram. Soc.*, **73**, 31–34 (1990).

430. Y. Yoshikawa and K. Tsuzuki, "Susceptibility to agglomeration of fine PLZT powders prepared from nitrate solutions," *J. Eur. Ceram. Soc.*, **6**, 227–35 (1990).

431. J. V. Biggers and S. Venkataramani, "Preparation and reactivity of lead zirconate-titanate solid solutions produced by precipitation from aqueous solutions," *Mater. Res. Bull.*, **13**, 717–22 (1978).

432. P. Duran and C. Moure, "Sintering at near theoretical density and properties of PZT ceramics chemically prepared," *J. Mater. Sci.*, **20**, 827–37 (1985).

433. H. Yamamura, M. Tanada, H. Haneda, S. Shirasaki and Y. Moriyoshi, "Preparation of PLZT by oxalate method in ethanol solution," *Ceram. Int.*, **11**, 23–26 (1985).

434. B.-M. Song, D.-Y. Kim and H. Yamamura, "Complete precipitation of PLZT oxalate by ammonia addition," *Ceram. Int.*, **12**, 179–80 (1986).

435. M. A. Akbas and W. E. Lee, "Synthesis and sintering of PLZT powder made by freeze/alcohol drying or gelation of citrate solutions," *J. Eur. Ceram. Soc.*, **15**, 57–63 (1995).

436. C. M. R. Bastos, M. Jafelicci, Jr., J. A. Varela and M. A. Zaghette, "Preparation of PZT powder from citrates"; pp. 1983–90 in *Ceramics Today–Tomorrow's Ceramics*. Edited by P. Vincenzini. Elsevier Science Publishers, Amsterdam, 1991.

437. H.-W. Wang, D. A. Hall and F. R. Sale, "Phase homogeneity and segregation in PZT powders prepared by thermal decomposition of metal-EDTA complexes derived from nitrate and chloride solutions," *J. Am. Ceram. Soc.*, **75**, 124–30 (1992).

438. T. R. N. Kutty and R. Balachandran, "Direct precipitation of lead zirconate titanate by the hydrothermal method," *Mater. Res. Bull.*, **19**, 1479–88 (1984).

439. K. C. Beal, "Precipitation of lead zirconate titanate solid solutions under hydrothermal conditions"; pp. 33-41 in *Adv. Ceram.*, **21** , *Ceramic Powder Science*. Edited by G. L. Messing, K. S. Mazdiyasni, J. W. McCauley and R. A. Haber. American Ceramic Society, Westerville, OH, 1987.

440. K. H. Lee, K. Asagsa, T. Ichihara and M. Daimon, "Synthesis of PZT crystalline powders by reaction of aqueous solution below 200°C," *Yogyo-Kyokai-Shi*, **95**, 736–40 (1987). [in Japanese]

441. H. Cheng, J. Ma, B. Zhu and Y. Cui, "Reaction mechanisms in the formation of lead zirconate titanate solid solutions under hydrothermal conditions," *J. Am. Ceram. Soc.*, **76**, 625–29 (1993).

442. Y. Ohba, T. Rikitoku, T. Tsurumi and M. Daimon, "Precipitation of lead zirconate titanate powders under hydrothermal conditions," *J. Ceram. Soc. Japan*, **104**, 6–10 (1996).

443. R. W. Schwartz, D. J. Eichorst and D. A. Payne, "Precipitation of PZT and PLZT powders using a continuous reactor"; pp. 123–28 in *Mater. Res. Soc. Symp. Proc.*, **73**, *Better Ceramics Through Chemistry II*. Edited by C. J. Brinker, D. E. Clark and D. R. Ulrich. Materials Research Society, Pittsburgh, PA, 1986.

444. L. M. Brown and K. S. Mazdiyasni, "Cold-pressing and low-temperature sintering of alkoxy-derived PLZT," *J. Am. Ceram. Soc.*, **55**, 541–44 (1972).

445. H. Hirashima, E. Onishi and M. Nakagawa, "Preparation of PZT powders from metal alkoxides," *J. Non-Cryst. Solids*, **121**, 404–406 (1990).

446. V. R. Palkar and M. S. Multani, "High-density PZT material," *Mater. Res. Bull.*, **14**, 1353–56 (1979).

447. G. Tomandl, A. Stiegelschmitt and R. Böhner, "Lowering the sintering temperature of PZT-ceramic by sol-gel processing"; pp. 305–308 in *Sci. Ceram.*, **14**. Edited by D. Taylor. The Institute of Ceramics, Stoke-on-Trent, U. K., 1988.

448. S. R. Gurkovich and J. B. Blum, "Preparation of monolithic lead titanate by a sol-gel process"; pp. 152–60 in *Ultrastructure Processing of Ceramics, Glasses, and Composites*. Edited by L. L. Hench and D. R. Ulrich. John Wiley, New York, 1984.

449. Z. Q. Zhuang, M. J. Haun, S.-J. Jang and L. E. Cross, "Fabrication and characterization of pure homogeneous PZT ceramics from sol-gel derived powders," *Adv. Ceram. Mater.*, **3**, 485–90 (1988).

450. R. Ostertag, G. Rinn, G. Tünker and H. Schmidt, "Preparation and properties of sol-gel-derived PZT powders"; pp. 11–20 in *Brit. Ceram. Proc.*, **41**, *Electroceramics*. Edited by A. J. Moulson, J. Binner and R. Morrell. The Institute of Ceramics, Stoke-on-Trent, U. K., 1989.

451. K. Hoselitz, "The magnetic properties of ferrites"; pp. 97–115 in *Brit. Ceram. Proc.*, **2**, *Magnetic Ceramics*. British Ceramic Society, Stoke-on-Trent, U. K., 1964.

452. T. Y. Tseng and J. C. Lin, "Preparation of fine-grained Ni-Zn ferrites," *J. Mater. Sci. Lett.*, **8**, 261–62 (1989).

453. T. Y. Tseng and S. Y. Jou, "Preparation of homogeneously grained Mn-Zn ferrites," *J. Mater. Sci. Lett.*, **8**, 777–78 (1989).

454. A. I. Busev, N. V. Kobrya, P. P. Korostelev and G. K. Mikhailov, "Production of ferrite powders by coprecipitation of 8-hydroxyquinolinates," *Inorg. Mater.*, **16**, 259–61 (1980).

455. P. K. Gallagher, H. M. O'bryan, Jr., F. Schrey and F. R. Monforte, "Preparation of a nickel ferrite from coprecipitated $Ni_{0.2}Fe_{0.8}C_2O_4.2H_2O$," *Am. Ceram. Soc. Bull.*, **48**, 1053–59 (1969).

456. M. Kiyama, T. Takada, N. Nagai and N. Horiishi, "Formation of hexagonal, platelike particles of $BaFe_{12-2x}Ti(IV)_xCo(II)_xO_{19}$ with 0 £ x < 1.2"; pp. 51–57 in *Adv. Ceram.*, **15**, *Fourth International Conference on Ferrites, Part I*. Edited by F. F. Y. Wang, American Ceramic Society, Columbus, OH, 1985.

457. A. Ataie, M. R. Piramoon, I. R. Harris and C. B. Ponton, "Effect of hydrothermal synthesis environment on the particle morphology, chemistry and magnetic properties of barium hexaferrite," *J. Mater. Sci.*, **30**, 5600–5606 (1995).

458. W. A. Kaczmarek, B. W. Ninham and A. Calka, "Structure and magnetic properties of aerosol synthesized barium ferrite particles," *J. Appl. Phys.*, **70**, 5909–5911 (1991).

459. M. V. Cabanas, J. M. Gonzalez-Calbet and M. Vallet-Regi, "Synthesis of barium hexaferrite by pyrolysis of an aerosol," *J. Mater. Res.*, **9**, 712–16 (1994).

460. M. V. Cabanas, J. M. Gonzalez-Calbet, M. Labeau, P. Mollard, M. Parnet and M. Vallet-Regi, "Evolution of the microstructure and its influence on the magnetic properties of aerosol synthesized $BaFe_{12}O_{19}$ particles," *J. Sold State Chemistry,* **101**, 265–74 (1992).

461. Y. Senzaki, J. Caruso, M. J. Hampden-Smith, T. T. Kodas and L.- M. Wang, "Preparation of strontium ferrite particles by spray pyrolysis," *J. Am. Ceram. Soc.*, **78**, 2973–76 (1995).

462. K. Matsumoto, Y. Yamanobe, S. Sasaki, T. Fujii, K. Honda and T. Miyamoto, "Preparation of YIG fine particles by mist pyrolysis," *J. Appl. Phys.*, **70**, 5912–14 (1991).

463. W. F. Kladnig and M. F. Zenger, "Preparation of fine-grained high-m ferrites by spray-roasting nitrate solutions," *J. Eur. Ceram. Soc.*, **9**, 341–49 (1992).

464. J. G. M. Delau, "Preparation of ceramic powders from sulphate solutions by spray drying and roasting," *Am. Ceram. Soc. Bull.*, **49**, 572–74 (1970).

465. B. J. Mulder, "Preparation of $BaTiO_3$ and other ceramic powders by coprecipitation of citrates in an alcohol," *Am. Ceram. Soc. Bull.*, **49**, 990–93 (1970).

466. T. J. Miller and Y. S. Kim, "Preparation of Ni, Zn-doped lithium ferrite by liquid drying," *Am. Ceram. Soc. Bull.*, **54**, 307–309 (1975).

467. S.-I. Hirano, J.-I. Watanabe and S. Naka, "Synthesis and properties of $CoFe_2O_4$ by hydrolysis of metal-acetylacetonates"; pp. 65–71 in *Adv. Ceram.*, **15**, *Fourth International Conference on Ferrites, Part I.* Edited by F. F. Y. Wang. American Ceramic Society, Columbus, OH, 1985.

468. K. Higuchi, S. Naka and S.-I. Hirano, "Synthesis and properties of Co^{2+}–Ti^{4+}-substituted barium hexaferrite by hydrolysis of organometallic compounds," *Adv. Ceram. Mater.*, **3**, 278–81 (1988).

469. S. Taketomi, K. Kawasaki, Y. Ozaki, S. Yuasa, Y. Otani and H. Miyajima, "Characteristics of yttrium iron garnet ultrafine particles prepared by the alkoxide method," *J. Am. Ceram. Soc.*, **77**, 1787–92 (1994).

470. P. Brahma, D. Chakravorty, K. Singh and D. Bahadur, "Preparation of barium hexagonal ferrites by sol-gel method," *J. Mater. Sci. Lett.*, **9**, 1438–40 (1990).

471. J.-H. Choy, Y.-S. Han and S.-W. Song, "Preparation and magnetic properties of

ultrafine $SrFe_{12}O_{19}$ particles derived from a metal citrate complex," *Mater. Lett.*, **19**, 257–62 (1994).

472. R. Chandrasekhar, S. W. Charles, K. O'grady, S. Morup and J. van Wonterghem, "Preparation and characterization of barium hexaferrite powders produced by decomposition of organometallic complexes," *Adv. Ceram. Mater.*, **2**, 65–68 (1987).

473. C. Marcilly, P. Courty and B. Delmon, "Preparation of highly dispersed mixed oxides and oxide solid solutions by pyrolysis of amorphous organic precursors," *J. Am. Ceram. Soc.*, **53**, 56–57 (1970).

474. K. Matsumoto, K. Yamaguchi, T. Fujii and A. Ueno, "Preparation of bismuth-substituted yttrium iron garnet powders by the citrate gel process," *J. Appl. Phys.*, **69**, 5918–20 (1991).

475. C. J. Chen, K. Bridger, S. R. Winzer and V. PaiVerneker, "A novel low-temperature preparation of Ni-Zn ferrite and the properites of the ultrafine particles formed," *J. Appl. Phys.*, **63**, 3786–88 (1988).

476. M. Ueda, S. Shimada and M. Inagaki, "Low temperature synthesis of zinc ferrite using hydrazine monohydrate," *J. Eur. Ceram. Soc.*, **15**, 265–69 (1995).

477. P. Rabindranathan and K. C. Patil, "A low temperature path to the preparation of ultrafine ferrites," *Am. Ceram. Soc. Bull.*, **66**, 688–92 (1987).

478. S. S. Manoharan and K. C. Patil, "Preparation and properties of fine particle Mg-Mn ferrite"; pp. 43–47 in *Adv. Ferrites*, **1**, *Proceedings of the Fifth International Conference on Ferrites*. Edited by C. M. Srivastava and M. J. Patni. Oxford and IBH Publishing Co., Pvt. Ltd., New Delhi, 1989.

479. A. Tsuzuki, K. Kani, K. Watari, S. Kawakami and Y. Torii, "Preparation of aluminium, gallium or indium-substituted $BaFe_{12}O_{19}$ powders by the glass-ceramic method," *J. Mater. Sci. Lett.*, **11**, 893–95 (1992).

480. K. Oda, T. Yoshio, K. O. Oka and F. Kanamaru, "Magnetic properties of $SrFe_{12}O_{19}$ particles prepared by the glass-ceramic method," *J. Mater. Sci. Lett.*, **3**, 1007–1010 (1984).

481. Y. Hayashi, T. Kimura and T. Yamaguchi, "Synthesis of acicular ferrite powders"; pp. 151–55 in *Adv. Ceram.*, **15**, *Fourth International Conference on Ferrites, Part I.* Edited by F. F. Y. Wang. American Ceramic Society, Columbus, OH, 1985.

482. K. H. Yoon, D. H. Lee, H. J. Jung and S. O. Yoon, "Molten salt synthesis of anisotropic $BaFe_{12}O_{19}$ powders," *J. Mater. Sci.*, **27**, 2941–45 (1992).

483. K. Higuchi, S. Takahashi, H. Itoh and S. Naka, "Synthesis of barium hexaferrite for magnetic recording media using the KCl flux system," *J. Mater. Sci.*, **23**, 588–92 (1988).

484. S. M. Gupta and A. R. Kulkarni, "Synthesis and dielectric properties of lead magnesium niobate—a review," *Mater. Chem. Phys.*, **39**, 98–109 (1994).

485. H. U. Anderson, M. J. Pennell and J. P. Guha, "Polymeric synthesis of lead magnesium niobate powders"; pp. 91–98 in *Adv. Ceram.*, **21**, *Ceramic Powder Science.* Edited by G. L. Messing, K. S. Mazdiyasni, J. W. McCauley and R.A. Haber. American Ceramic Society, Westerville, OH, 1987.

486. S. L. Swartz and T. R. Shrout, "Fabrication of perovskite lead magnesium niobate," *Mater. Res. Bull.*, **17**, 1245–50 (1982).

487. M. Lejeune and J. P. Boilot, "Optimization of dielectric properties of lead-magnesium niobate ceramics," *Am. Ceram. Soc. Bull.*, **65**, 679–82 (1986).

488. M. Lejeune and J. P. Boilot, "Low firing dieletrics based on lead magnesium niobate," *Mater. Res. Bull.*, **20**, 493–99 (1985).

489. M. Lejeune and J. P. Boilot, "Formation mechanism and ceramic process of the ferroelectric perovskites : $Pb(Mg_{1/3}Nb_{2/3})O_3$ and $Pb(Fe_{1/2}Nb_{1/2})O_3$," *Ceram. Int.*, **8**, 99–103 (1982).

490. S. L. Swartz, T. R. Shrout, W. A. Schulze and L. E. Cross, "Dielectic properties of lead-magnesium niobate ceramics," *J. Amer. Ceram. Soc.*, **67**, 311–15 (1984).

491. D. Saha, A. Sen and H. S. Maiti, "Solid-state synthesis of precursor $MgNb_2O_6$ for the preparation of $Pb(Mg_{1/3}Nb_{2/3})O_3$," *J. Mater. Sci. Lett.*, **13**, 723–24 (1994).

492. J. P. Guha and H. U. Anderson, "Preparation of perovskite $Pb(Mg_{1/3}Nb_{2/3})O_3$ using $Pb_3Nb_2O_8$ and MgO," *J. Amer. Ceram. Soc.*, **69**, C-287–C-288 (1986).

493. A. Watanabe, H. Haneda, Y. Moriyoshi, S. Shirasaki, S. Kuramoto and H. Yamamura, "Praparation of lead magnesium niobate by a coprecipitation method," *J. Mater. Sci.*, **27**, 1245–49 (1992).

494. P. Ravindranathan, S. Komarneni, A. S. Bhalla, R. Roy and L. E. Cross, "Sol-gel processing of lead magnesium niobate (PMN) powder and its characterization"; pp. 182–89 in *Ceram. Trans.* **1**, *Ceramic Powder Science II*. Edited by G. L. Messing, E. R. Fuller, Jr. and H. Hausner. American Ceramic Society, Westerville, OH, 1988.

495. T. Fukui, C. Sakurai and M. Okuyama, "Lower-temperature preparation of $Pb\ (Mg_{1/3}Nb_{2/3})O_3$ with a perovskite structure by the complex alkoxide method," *J. Non-Cryst. Solids*, **134**, 293–95 (1991).

496. F. Chaput, J.-P. Boilot, M. Lejeune, R. Papiernik and L. G. Hubert-Pfalzgraf, "Low-temperature route to lead magnesium niobate," *J. Amer. Ceram. Soc.*, **72**, 1355–57 (1989).

497. J.-H. Choy, J.-S. Yoo, S.-G. Kang, S.-T. Hong and D.-G. Kim, "Ultra-fine $Pb(Mg_{1/3}Nb_{2/3})O_3$ (PMN) powder synthesized from metal-citrate gel by thermal shock method," *Mater. Res. Bull.*, **25**, 283–91 (1990).

498. S. M. Gupta and A. R. Kulkarni, "Dielectric and microstructure studies of lead magnesium niobate prepared by partial oxalate route," *J. Eur. Ceram. Soc.*, **16**, 473–80 (1996).

499. C. N. R. Rao, R. Nagarajan and R. Vijayaraghavan, "Synthesis of cuprate superconductors," *Supercond. Sci. Technol.*, **6**, 1–22 (1993).

500. X.-D. Chen, S. Y. Lee, J. P. Golben, S.-I. Lee, R. D. McMichael, Y. Song, T. W. Noh and J. R. Gaines, "Practical preparation of copper oxide superconductors," *Rev. Sci. Instruments*, **58**, 1565–71 (1987).

501. J. R. Spann, I. K. Lloyd, M. Kahn and M. T. Chase, "Preparation of orthorhombic $Ba_2YCu_3O_7$ powder by single-step calcining," *J. Am. Ceram. Soc.*, **73**, 435–38 (1990).

502. H. C. Ling, "Preparation and densification of superconducting $YBa_2Cu_3O_x$ ceramics," *J. Mater. Sci.*, **25**, 3297–3308 (1990).

503. J. Tartaj, C. Moure, P. Duran, J. L. Garcia-Fierro and J. Colino, "Processing and properties of superconducting $YBa_2Cu_3O_{7-x}$ powder by single-step calcining in air," *J. Mater. Sci.*, **26**, 6135–43 (1991).

504. Y. Sakka and M. Ohtaguchi, "Characterization of Bi-Pb-Sr-Ca-Cu-O oxide powders synthesized by the oxalate coprecipitation route," *J. Mater. Sci. Lett.*, **11**, 749–53 (1992).

505. P. Pramanik, B. K. Raul, S. Chakrabarti, S. Biswas, D. Bhattacharya and K. L. Chopra, "Corprecipitation route to ceramic superconductors"; pp. 241–53 in *Studies of High Temperature Superconductors,* Vol. 2. Edited by A. Narlikar. Nova Science Publishers, Commack, 1989.

506. C.-T. Chu and B. Dunn, "Preparation of high-T_c superconducting oxides by the amorphous citrate process," *J. Am. Ceram. Soc.*, **70**, C-375–C-377 (1987).

507. C. S. Niou, Y. T. Ma, W. P. Li, J. Javadpour and L. E. Murr, "Preparation of superconducting $YBa_2Cu_3O_{7-x}$ powders by a solution technique," *J. Mater. Sci. Mater. Electorn.*, **3**, 181–86 (1992).

508. P. Karen and A. Kjekshus, "Citrate-gel syntheses in the Y(O)-Ba(O)-Cu(O) system," *J. Am. Ceram. Soc.*, **77**, 547–52 (1994).

509. I. Sargankova, M. Timko, J. Kovac, S. Mat'as, P. Diko and M. Cernik, "Synthesis of Y-Ba-Cu-O fine powder and some of its magnetic properties," *J. Mater. Sci. Lett.*, **11**, 1718–20 (1992).

510. P. Sujatha Devi and H. S. Maiti, "A novel autoignited combustion process for the synthesis of Bi-Pb-Sr-Ca-Cu-O superconductors with a $T_c(O)$ of 125 K," *J. Solid State Chem.*, **109**, 35–42 (1994).

511. W. J. Thomson, H. Wang, D. B. Parkman, D. X. Li, M. Strasik, T. S. Luhman, C. Han and I. A. Aksay, "Reaction sequencing during processing of the 123 superconductor," *J. Am. Ceram. Soc.*, **72**, 1977–79 (1989).

512. R. T. Richardson and C. D. Veitch, "Synthesis of bismuth/strontium/calcium/copper and yttrium/barium/copper oxide systems from malonate precursors," *J. Mater. Sci.*, **29**, 1487–91 (1994).

513. A. Pebler and R. G. Charles, "Synthesis of small particle size $YBa_2Cu_3O_{7-x}$ by a vapour phase process," *Mater. Res. Bull.*, **23**, 1337–44 (1988).

514. I.-T. Kim, T.-S. Oh and Y.-H. Kim, "Preparation and characterization of $YBa_2Cu_3O_{7-x}$ powder by the ultrasonic spray pyrolysis method," *J. Mater. Sci.*, **26**, 6275–78 (1991).

515. T. T. Kodas, E. M. Engler, V. Y. Lee, R. Jacowitz, T. H. Baum, K. Roche and S. S. P. Parkin, "Aerosol flow reactor production of fine $Y_1Ba_2Cu_3O_7$ powder : fabrication of superconducting ceramics," *Appl. Phys. Lett.*, **52**, 1622–24 (1988).

516. N. Tohge, M. Tatsumisago, T. Minami, K. Okuyama, K. Arai, Y. Inada and Y. Kousaka, " Preparation of superconducting fine particles in the Bi-(Pb)-Ca-Sr-Cu-O system using the spray-pyrolysis method," *J. Mater. Sci. Mater. Electron.*, **1**, 46–48 (1990).

517. N. Tohge, M. Tatsumisago, T. Minami, M. Adachi, Y. Kousaka and K. Okuyama, "Preparation conditions and morphology of superconducting fine particles in the Bi-Ca-Sr-Cu-O system prepared by spray pyrolysis," *J. Am. Ceram. Soc.*, **74**, 2117–22 (1991).

518. S. Katayama and M. Sekine, "Superconducting oxide $YBa_2Cu_3O_{7-x}$ prepared by the metal alkoxide method," *J. Mater. Res.*, **5**, 683–90 (1990).

519. R. Sivakumar, "Ceramic materials as implants," *Trans. Ind. Ceram. Soc.*, **54**, 33–47 (1995).

520. M. Asada, K. Oukami, S. Nakamura and K. Takahashi, "Effect of powder characteristics on the sinterability of calcium hydroxyapatite," *Yogyo-Kyokai-Shi*, **95**, 703–709 (1987). [in Japanese]

521. M. Jarcho, C. H. Bolen, M. B. Thomas, J. Bobick, J. F. Kay and R. H. Doremus, "Hydroxyapatite synthesis and characterization in dense polycrystalline form," *J. Mater. Sci.*, **11**, 2027–35 (1976).

522. Z. Bako and I. Kotsis, "Composition of precipitated calcium phosphate ceramics," *Ceram. Int.*, **18**, 373–78 (1992).

523. M. Aizawa, K. Itatani, F. S. Howell and A. Kishioka, "Effects of starting materials on properties of hydroxyapatite powders by spray-pyrolysis technique," *J. Ceram. Soc. Japan*, **104**, 126–32 (1996). [in Japanese]

524. S. Inoue and A. Ono, "Preparation of hydroxyapatite by spray-pyrolysis technique," *Yogyo-Kyokai-Shi*, **95**, 759–63 (1987).

525. T. Hattori, Y. Iwadate, H. Inai, K. Sato and Y. Imai, "Preparation of hydroxyapatite powder using a freeze-drying method," *Yogyo-Kyokai-Shi*, **95**, 825–27 (1987).

526. A. Deptula, W. Lada, T. Olczak, A. Borello, C. Alvani and A. di Bartolomeo, "Preparation of spherical powders of hydroxyapatite by sol-gel process," *J. Non-Cryst. Solids*, **147 & 148**, 537–41 (1992).

527. A. Osaka, Y. Miura, K. Takeuchi, M. Asada and K. Takahashi, "Calcium apatite prepared from calcium hydroxide and orthophosphoric acid," *J. Mater. Sci. Mater. Medicine*, **2**, 51–55 (1991).

528. M. Akao, H. Aoki and K. Kato, "Mechanical properties of sintered hydroxyapatite for prosthetic applications," *J. Mater. Sci.*, **16**, 809–812 (1981).

529. T. Goto, N. Wakamatsu, H. Kamemizu, M. Iijima, Y. Doi and Y. Moriwaki, "Sintering mechanism of hydroxyapatite by addition of lithium phosphate," *J. Mater. Sci. Mater. Medicine*, **2**, 149–52 (1991).

530. F. Yamada, Y. Kaneko and H. Iwasaki, "Synthetic condition for calcium phytate to obtain hydroxyapatite powder," *Nippon Kagaku Kaishi*, 1712–17 (1989). [in Japanese].

531. M. Aizawa, K. Itatani, F. S. Howell and A. Kishioka, "Some properties of carbonate-containing hydroxyapatite powder prepared by spray-pyrolysis technique using urea as a foaming agent," *J. Ceram. Soc. Japan.*, **103**, 1214–19 (1995).

532. M. Toriyama, A. Ravaglioli, A. Krajewski, G. Celotti and A. Piancastelli, "Synthesis of hydroxy-apatite-based powders by mechano-chemical method and their sintering," *J. Eur. Ceram. Soc.*, **16**, 429–36 (1996).

533. V. I. Matkovich and P. M. Corbett, "Formation of zircon from zirconium dioxide and silicon dioxide in the presence of vanadium pentoxide," *J. Am. Ceram. Soc.*, **44**, 128–30 (1961).

534. T. Demiray, D. K. Nath and F. A. Hummel, "Zircon-vanadium blue pigment," *J. Am. Ceram. Soc.*, **53**, 1–4 (1970).

535. Y. Kadogawa and T. Yamate, "Synthesis of zircon by the sol-gel method," *Yogyo-Kyokai-Shi*, **93**, 338–40 (1985). [in Japanese]

536. G. Monros, J. Carda, M. A. Tena, P. Escribano, M. Sales and J. Alarcon, "Different kinds of solid solutions in the V_2O_5-$ZrSiO_4$-NaF system by sol-gel processes and their characterization," *J. Eur. Ceram. Soc.*, **11**, 77–86 (1993).

537. G. Vilmin, S. Komarneni and R. Roy, "Lowering crystallization temperature of zircon by nanoheterogeneous sol-gel processing," *J. Mater. Sci.*, **22**, 3556–60 (1987).

538. Y. Kanno, "Thermodynamic and crystallographic discussion of the formation and dissociation of zircon," *J. Mater. Sci.*, **24**, 2415–20 (1989).

539. T. Mori, H. Hoshino, Y. Ishikawa, T. Yamaguchi, H. Yamamura, H. Kobayashi and T. Mitamura, "Preparation of $ZrSiO_4$ powder using sol-gel process, Part 4. Preparation from $ZrOCl_2$ and colloidal SiO_2," *J. Ceram. Soc. Japan.*, **99**, 227–32 (1991). [in Japanese]

540. R. F. Haaker and R. C. Ewing, "Solution-gelation method for preparing polycrystalline zircon," *J. Am. Ceram. Soc.*, **64**, C-149 (1981).

541. H. Kobayashi, T. Terasaki, T. Mori, H. Yamamura and T. Mitamura, "Preparation of $ZrSiO_4$ powders by sol-gel process, Part 3. Preparation conditions of $ZrSiO_4$ composition precursor gels from $Si(OC_2H_5)_4$ and $Zr(O^iC_3H_7)_4$ alkoxides," *J. Ceram. Soc. Japan.*, **99**, 42–46 (1991). [in Japanese]

542. A. B. Hardy and W. E. Rhine, "Preparation of zircon and mullite-zircon powders by sol-gel techniques"; pp. 577–84 in *Chemical Processing of Advanced Materials.* Edited by L. L. Hench and J. K. West, John Wiley, New York, 1992.

543. H. Kido and S. Komarneni, "Hydrothermal processing of zircon," *Trans. Mater. Res. Soc. Japan.*, 358–69 (1990).

544. M. D. Karkhanavala and F. A. Hummel, "The polymorphism of cordierite," *J. Am. Ceram. Soc.*, **36**, 389–92 (1953).

545. G. V. Gibbs, "The polymorphism of cordierite I : the crystal structure of low cordierite," *Am. Min.*, **51**, 1068–87 (1966).

546. A. M. Kazakos, S. Komarneni and R. Roy, "Sol-gel processing of cordierite : effect of seeding and optimization of heat treatment," *J. Mater. Res.*, **5**, 1095–1103 (1990).

547. M. Okuyama, T. Fukui and C. Sakurai, "Effects of complex precursors on alkoxide-derived cordierite powder," *J. Am. Ceram. Soc.*, **75**, 153–60 (1992).

548. M. Okuyama, T. Fukui and C. Sakurai, "Effects of solvent and alkoxy group on powder precipitation of cordierite by complex-alkoxide hydrolysis," *J. Non-Cryst. Solids.*, **143**, 112–120 (1992).

549. M. Okuyama, T. Fukui and C. Sakurai, "Effects of addition rate and pH of water on

cordierite powder prepared by complex-alkoxide hydrolysis," *J. Non-Cryst. Solids.*, **144**, 298–304 (1992).

550. H. Suzuki, K. Ota and H. Saito, "Preparation of cordierite ceramics from metal alkoxides (Part 1); preparation and characterization of the powder," *Yogyo-Kyokai-Shi.*, **95**, 163–69 (1987).

551. C. Gensse and U. Chowdhry, "Non-conventional route to glass-ceramics for electronic packaging"; pp. 693–703 in *Mater. Res. Soc. Symp. Proc.*, **73**, *Better Ceramics Through Chemistry II.* Edited by C. J. Brinker, D. E. Clark and D. R. Ulrich. Materials Reserch Society, Pittsburgh, PA, 1986.

552. R. Gopi Chandran and K. C. Patil, "Combustion synthesis, characterization, sintering and microstructure of cordierite," *Brit. Ceram. Trans.*, **92**, 239–45 (1993).

553. C. Ganguly, "Advanced methods for the fabrication of mixed uranium, plutonium oxide, monocarbide and mononitride fuels for fast breeder reactors," *Trans. Ind. Ceram. Soc.*, **47**, 161–71 (1988).

554. R. G. Wymer and J. H. Coobs, "Preparation, coating, evaluation and irradiation testing of sol-gel oxide microspheres"; pp. 61–79 in *Brit. Ceram. Proc.*, **7**, *Nuclear and Engineering Ceramics.* British Ceramic Society, Stoke-on-Trent, U. K., 1967.

555. D. Segal, *Chemical Synthesis of Advanced Ceramic Materials*; pp. 42–46. Cambridge University Press, Cambridge, 1989.

556. E. Zimmer, C. Ganguly, J. Borchardt and H. Langen, "SGMP – an advanced method for fabrication of UO_2 and MOX fuel pellets," *J. Nucl. Mater.*, **152**, 169–77 (1988).

557. T. Yamada, T. Kawahito and T. Iwai, "Crystallization of amorphous Si_3N_4 prepared by the thermal decomposition of $Si(NH)_2$," *J. Mater. Sci. Lett.*, **2**, 275–78 (1983).

558. C. Qitao, M. G. Zadnik, S. Gloggner, L. Deyu and B. H. O'Connor, "Characteristics of silicon nitride powder synthesized from Western Australian silicon"; pp. 604–609 in *Key Eng. Mater.*, **53–55**, *Austceram '90.* Edited by P. J. Darragh and R. J. Stead. Trans Tech Publications, Switzerland, 1991.

559. K. Hirao, Y. Miyamoto and M. Koizumi, "Combustion synthesis of nitride powders under high nitrogen pressure"; pp. 289–300 in *Adv. Ceram.*, **21**, *Ceramic Powder Science.* Edited by G. L. Messing, K. S. Mazdiyasni, J. W. McCauley and R. A. Haber. American Ceramic Society, Westerville, OH, 1987.

560. H. J. Lee, K. Eguchi and T. Yoshida, "Preparation of ultrafine silicon nitride, and silicon nitride and silicon carbide mixed powders in a hybrid plasma," *J. Amer. Ceram. Soc.*, **73**, 3356–62 (1990).

561. F. Allaire and S. Dallaire, "Synthesis and characterization of silicon nitride powders produced in a d.c. thermal plasma reactor," *J. Mater. Sci.*, **26**, 6736–40 (1991).

562. S. Prochazka and C. Greskovich, "Synthesis and characterization of a pure silicon nitride powder," *Am. Ceram. Soc. Bull.*, **57**, 579–81 (1978).

563. J. Janiga, K. P. Sin and V. Figusch, "Synthesis of silicon nitride powder by gas-phase reaction," *J. Eur. Ceram. Soc.*, **8**, 153–60 (1991).

564. S.-C. Zhang and W. R. Cannon, "Preparation of silicon nitride from silica," *J. Amer. Ceram. Soc.*, **67**, 691–95 (1984).

565. W. M. Shen and C. F. Chang, "Fine Si_3N_4 and SiC powders prepared by vapour-phase pyrolysis"; pp. 193–201 in *Adv. Ceram.*, **21**, *Ceramic Powder Science*. Edited by G. L. Messing, K. S. Mazdiyasni, J. W. McCauley and R. A. Haber. American Ceramic Society, Westerville, OH, 1987.

566. M. Luce, N. Herlin, O. Croix and M. Cauchetier, "Nanometric Si-based powders formed by aerosol-laser interaction," *Silic. Ind.*, **60**, 77–81 (1995).

567. T. J. Mroz, Jr. "Aluminium nitride," *Am. Ceram. Soc. Bull.*, **72** (6), 78–80 (1993).

568. French Patent 91, 11801 (September, 1991).

569. P. Lefort and M. Billy, "Mechanism of AlN formation through the carbothermal reduction of Al_2O_3 in a flowing N_2 atmosphere," *J. Am. Ceram. Soc.*, **76**, 2295–99 (1993).

570. U. Nadkarni, "Non-oxide ceramic powder"; pp. 81–86 in *Brit. Ceram. Proc.*, **47**, *Fine Ceramic Powders*. Edited by R. Freer and J. L. Woodhead. The Institute of Ceramic, Stoke-on-Trent, U. K., 1991.

571. S. E. Pratsinis, G. Wang, S. Panda, T. Guiton and A. W. Weimer, "Aerosol synthesis of AlN by nitridation of aluminium vapour and clusters," *J. Mater. Res.*, **10**, 512–20 (1995).

572. K. Itatani, K. Sano, F. S. Howell, A. Kishioka and M. Kinoshita, "Some properties of aluminium nitride powder synthesized by low-pressure chemical vapour deposition," *J. Mater. Sci.*, **28**, 1631–38 (1993).

573. K. Baba, N. Shohata and M. Yonezawa, "Synthesis and properties of ultrafine AlN powder by rf plasma," *Appl. Phys. Lett.*, **54**, 2309–2311 (1989).

574. Z.-P. Lu and E. Pfender, "DC plasma synthesis of aluminium nitride ceramic powders"; pp. 857–60 in *Mater. Res. Soc. Symp. Proc.*, **180**, *Better Ceramics Through Chemistry IV.* Edited by B. J. J. Zelinski, C. J. Brinker, D. E. Clark and D. R. Ulrich. Materials Research Society, Pittsburgh, PA, 1990.

575. A. Tsuge, H. Inoue, M. Kasori and K. Shinozaki, "Raw material effect on AlN powder synthesis from Al_2O_3 carbothermal reduction," *J. Mater. Sci.*, **25**, 2359–61 (1990).

576. B. Forslund and J. Zheng, "Synthesis of AlN powder by carbothermal nitridation of Al_2O_3 at elevated N_2 pressures"; pp. 213–20 in *Int. Symp. Ceram. Mater. Compon. Engines 4th Proc.*, **4**. Edited by R. Carlsson, T. Johansson and L. Kahlman, 1992.

577. Y.-W. Cho and J. A. Charles, "Synthesis of nitrogen ceramic powders by carbothermal reduction and nitridation : III, aluminium nitride," *Mater. Sci. Tech.*, **7**, 495–504 (1991).

578. B.-H. Kim, Y.-S. Jun and K. Hong, "Synthesis of AlN powder from alumina gel containing residual organic matter entrapped by sol-gel method"; pp. 595–601 in *Chemical Processing of Advanced Materials*. Edited by L. L. Hench and J. K. West, John Wiley, New York, 1992.

579. L. D. Silverman, "Carbothermal synthesis of aluminium nitride," *Adv. Ceram. Mater.*, **3**, 418–19 (1988).

580. K. G. Nickel, R. Riedel and G. Petzow, "Thermodynamic and experimental study of high-purity aluminium nitride formation from aluminium chloride by chemical vapour deposition," *J. Am. Ceram. Soc.*, **72**, 1804–1810 (1989).

581. R. Riedel and K.-U. Gaudl, "Formation and characterization of amorphous aluminium nitride powder and transparent aluminium nitride film by chemical vapour deposition," *J. Am. Ceram. Soc.*, **74**, 1331–34 (1991).

582. I. Kimura, N. Hotta, H. Nukui, N. Saito and S. Yasukawa, "Synthesis of fine AlN powder by vapour-phase reaction," *J. Mater. Sci. Lett.*, **7**, 66–68 (1988).

583. P. Greil, M. Kulig, D. Hotza, H. Lange and R. Tischtau, "Aluminium nitride ceramics with high thermal conductivity from gas-phase synthesized powders," *J. Eur. Ceram. Soc.*, **13**, 229–37 (1994).

584. J.-R. Park, S.-W. Rhee and K.-H. Lee, "Gas-phase synthesis of AlN powders from $AlCl_3$-NH_3-N_2," *J. Mater. Sci.*, **28**, 57–64 (1993).

585. Y. Sugahara, T. Onuma, O. Tanegashima, K. Kuroda and C. Kato, "Preparation of aluminium nitride from poly(isopropyliminoalane)," *J. Ceram. Soc. Japan.*, **100**, 101–103 (1992).

586. P. R. Coffman, W. T. Petuskey and S. K. Dey, "Chemical synthesis of aluminium nitride powders"; pp. 709–714 in *Mater. Res. Soc. Symp. Proc.*, **180**, *Better Ceramics Through Chemistry IV.* Edited by B. J. J. Zelinski, C. J. Brinker, D. E. Clark and D. R. Ulrich. Materials Research Society, Pittsburgh, PA, 1990.

587. K. Tsuchida, Y. Takeshita, A. Yamane and A. Kato, "Preparation of AlN powders by vapour phase reaction of $Al(i\text{-}Bu)_3$-NH_3 system," *Yogyo-Kyokai-Shi*, **95**, 1198–1201 (1987). [in Japanese]

588. A. A. Adjaottor and G. L. Griffin, "Aerosol synthesis of aluminium nitride powder using metalorganic reactants," *J. Am. Ceram. Soc.*, **75**, 3209–3214 (1992).

589. K. Kubo, K. Itatani, F. S. Howell, A. Kishioka and M. Kinoshita, "Some properties of aluminium-nitride powder prepared by metal-organic chemical vapour deposition," *J. Eur. Ceram. Soc.*, **15**, 661–66 (1995).

590. L. V. Interrante, L. E. Carpenter II, C. Whitmarsh, W. Lee, M. Garbauskas and G. A. Slack, "Studies of organometallic precursors to aluminium nitride"; pp. 359–66 in *Mater. Res. Soc. Symp. Proc.*, **73**, *Better Ceramics Through Chemistry II.* Edited by C. J. Brinker, D. E. Clark and D. R. Ulrich, Materials Research Society, Pittsburgh, PA, 1986.

591. I. C. Huseby, "Synthesis and characterization of a high-purity AlN powder," *J. Am. Ceram. Soc.*, **66**, 217–20 (1983).

592. S.-I. Hirano, T. Yogo, S. Asada and S. Naka, "Synthesis of amorphous boron nitride by pressure pyrolysis of borazine," *J. Am. Ceram. Soc.*, **72**, 66–70 (1989).

593. T. Yogo and S. Naka, "Synthesis of boron nitride from triammoniadecaborane and hydrazine under pressure," *J. Mater. Sci.*, **25**, 374–78 (1990).

594. V. L. Solozhenko and N. F. Ostrovskaya, "Structural peculiarities of graphite-like boron nitride produced by crystallization in the region of metastability," *Mater. Lett.*, **25**, 133–37 (1995).

595. R. H. Wentorf, Jr., "Cubic form of boron nitride," *J. Chem. Phys.*, **26**, 956 (1957).

596. S.-I. Hirano, T. Yamaguchi and S. Naka, "Effects of AlN additions and atmosphere on the synthesis of cubic boron nitride," *J. Am. Ceram. Soc.*, **64**, 734–36 (1981).

597. M. Wakatsuki, K. Ichinose and T. Aoki, "Synthesis of polycrystalline cubic BN," *Mater. Res. Bull.*, **7**, 999–1004 (1972).

598. K. Susa, T. Kobayashi and S. Taniguchi, "Catalytic effect of water on the synthesis of cubic BN," *Mater. Res. Bull.*, **9**, 1443–46 (1974).

599. E. Rapoport and S. Nadiv, "Mechanochemical activation of hexagonal boron nitride and synthesis of the cubic form," *J. Mater. Sci. Lett.*, **4**, 34–36 (1985).

600. T. Kobayashi, K. Susa and S. Taniguchi, "New catalysts for the high pressure synthesis of cubic BN," *Mater. Res. Bull.*, **10**, 1231–36 (1975).

601. R. Liepins, K. P. Staudhammer, K. A. Johnson and M. Thomson, "Shock-induced synthesis. I. Cubic boron nitride from ammonia borane," *Mater. Lett.*, **7**, 44–46 (1988).

602. U. S. Patent 492, 767 (1893).

603. N. N. Ault and J. T. Crowe, "Silicon carbide," *Am. Ceram. Soc. Bull.*, **72** (6), 114, 116 (1993).

604. N. N. Ault, "Raw materials for refractories : SiC and Si_3N_4," *Ceram. Eng. Sci. Proc.*, **4**, 186–93 (1983).

605. K. A. Schwetz, "Silicon carbide and its high-technology ceramics," *Radex Rundschau.*, 26–39 (1989).

606. G.C.-T. Wei, "Beta SiC powders produced by carbothermic reduction of silica in a high-temperature rotary furnace," *J. Am. Ceram. Soc.*, **66**, C-111–C-113 (1983).

607. G. C. Wei, C. R. Kennedy and L. A. Harris, "Synthesis of sinterable SiC powders by carbothermic reduction of gel-derived precursors and pyrolysis of polycarbosilane," *Am. Ceram. Soc. Bull.*, **63**, 1054–61 (1984).

608. H.-P. Martin, E. Müller, Y. Knoll, R. Strienitz and G. Schuster, "Silicon carbide derived from silica sol and sugar," *J. Mater. Sci. Lett.*, **14**, 620–22 (1995).

609. Lj. Cerovic, S. K. Milonjic and S. P. Zec, "A comparison of sol-gel derived silicon carbide powders from saccharose and activated carbon," *Ceram. Int.* **21**, 271–76 (1995).

610. H. Tanaka and Y. Kurachi, "Synthesis of b-SiC powder from organic precursor and its sinter-ability," *Ceram. Int.*, **14**, 109–115 (1988).

611. P. C. Kong and E. Pfender, "Formation of ultrafine b-silicon carbide powders in an argon thermal plasma jet," *Langmuir*, **3**, 259–65 (1987).

612. A. Kuibira, A. Mitsui, J. Hojo and A. Kato, "Preparation of SiC and Si_3N_4 powders by RF-plasma," *Yogyo-Kyokai-Shi*, **95**, 89–93 (1987). [in Japanese]

613. F. Allaire, L. Parent and S. Dallaire, "Production of submicron SiC particles by d. c. thermal plasma: a systematic approach based on injection parameters," *J. Mater. Sci.*, **26**, 4160–65 (1991).

614. J.-M. Lihrmann and M. Cauchetier, "A model for the formation of nanosized SiC powders by laser-induced gas-phase reaction," *J. Eur. Ceram. Soc.*, **13**, 41–46 (1994).

615. C. M. Hollabaugh, D. E. Hull, L. R. Newkirk and J. J. Petrovic, "RF plasma synthesis of ultrafine, ultrapure silicon carbide powder"; pp. 367–73 in *Ultrastructure Processing of Ceramics, Glasses, and Composites.* Edited by L. L. Hench and D. R. Ulrich. John Wiley, New York, 1984.

616. J. Kondo and G. Saiki, "Synthesis and sintering of boron doped SiC powders by plasma arc method"; pp. 285–92 in *Ceram. Trans.*, **1**, *Ceramic Powder Science II.* Edited by G. L. Messing, E. R. Fuller, Jr. and H. Hausner. American Ceramic Society, Westerville, OH, 1988.

617. Y. Okabe, J. Hojo and A. Kato, "Formation of fine silicon carbide powders by a vapour phase method," *J. Less-Common Metals.*, **68**, 29–41 (1979).

618. Y. Okabe, J. Hojo and A. Kato, "Formation of silicon carbide powders by the vapour phase reaction of the SiH_4-CH_4-H_2 system," *Nippon Kagaku Kaishi*, 188–193 (1980). [in Japanese]

619. Y. Okabe, K. Miyachi, J. Hojo and A. Kato, "Sintering behaviour of ultrafine silicon carbide powders obtained by a vapour phase reaction," *Nippon Kagaku Kaishi*, 1363–70 (1981). [in Japanese]

620. J. Hojo, K. Miyachi, Y. Okabe and A. Kato, "Effect of chemical composition on the sinterability of ultrafine SiC powders," *J. Am. Ceram. Soc.*, **66**, C-114–C-115 (1983).

621. W. Böcker and H. Hausner, "Preparation of silicon carbide powders by vapour phase reaction," *Ber. Deutsch, Keram. Gesell.*, **55**, 233–37 (1978) [in German]

622. M. Endo, T. Sano, K. Mori, N. Urasato and M. Shiraishi, "Preparation of ultrafine SiC powders by pyrolysis of tetramethyldisilane," *Yogyo-Kyokai-Shi*, **95**, 104–110 (1987). [in Japanese]

623. J. Y. Guo, F. Gitzhofer and M. I. Boulos, "Induction plasma synthesis of ultrafine SiC powders from silicon and CH_4," *J. Mater. Sci.*, **30**, 5589–99 (1995).

624. Y. Ando and R. Uyeda, "Preparation of ultrafine SiC particles by gas evaporation," *J. Am. Ceram. Soc.*, **64**, C-12–C-13 (1981).

625. M. T. Spohn, "Boron carbide," *Am. Ceram. Soc. Bull.*, **72** (6), 88–89 (1993).

626. F. Thevenot, "A review on boron carbide"; pp. 59–88 in *Key Eng. Mater.*, **56–57**, *Advanced Ceramics.* Edited by C. Ganguly, S. K. Roy and P. R. Roy. Trans Tech Publications, Zurich, 1991.

627. A. W. Weimer, W. G. Moore, R. P. Roach, J. E. Hitt, R. S. Dixit and S. E. Pratsinis, "Kinetics of carbothermal reduction synthesis of boron carbide," *J. Am. Ceram. Soc.*, **75**, 2509–2514 (1992).

628. J. B. Holt, "Synthesis of nonoxide ceramic powders by solid combustion processes"; pp. 301–310 in *Adv. Ceram.*, **21**, *Ceramic Powder Science.* Edited by G. L. Messing, K. S. Mazdiyasni, J. W. McCauley and R. A. Haber. American Ceramic Society, Westerville, OH, 1987.

629. A. K. Knudsen, "Laser-driven synthesis and densification of ultrafine boron carbide powders"; pp. 237–47 in *Adv. Ceram.*, **21**, *Ceramic Powder Science.* Edited by G. L. Messing, K. S. Mazdiyasni, J. W. McCauley and R. A. Haber. American Ceramic Society, Westerville, OH, 1987.

630. T. Ikeda, T. Mori, F. Noguchi, T. Iida and T. Mitamura, "Preparation of ultrafine zirconium nitride powders from zirconia by reduction with magnesium," *Yogyo-Kyokai-Shi*, **93**, 505–510 (1985). [in Japanese]

631. H. Kobayashi, M. Katoh, Y. Kamiyama and T. Mitamura, "Preparation of ZrN fine powders from ZrO_2 by reduction with Mg and their recovery," *J. Ceram. Soc. Japan.*, **99**, 508–513 (1991). [in Japanese]

632. Y. Kamiyama, T. Arakawa, T. Mori, T. Iida and T. Mitamura, "Preparation of zirconium nitride powder from zirconium chloride (IV)," *Yogyo-Kyokai-Shi.*, **92**, 29–34 (1984). [in Japanese]

633. A. N. Rubtsov, O. G. Kiselev, S. P. Kucherin and S. M. Rekkonen, "Investigation of the processes and products of nitriding of titanium and its hydride," *Inorg. Mater.*, **12**, 796–97 (1976).

634. R. P. Singh and R. D. Doherty, "Synthesis of titanium nitride powders under glow discharge plasma," *Mater. Lett.*, **9**, 87–89 (1990).

635. M. Yoshimura, M. Nishioka and S. Somiya, "Synthesis of TiN and TiC powders by a reduction/nitridation method using arc image heating," *J. Mater. Sci. Lett.*, **6**, 1463–65 (1987).

636. H. Kobayashi, K. Shimosaka, Y. Kaneda and T. Mitamura, "Preparation and formation processes of ZrC fine powders by Mg-thermite method," *J. Ceram. Soc. Japan.*, **101**, 190–94 (1993). [in Japanese].

637. T. Ikeda, T. Mori, T. Iida and T. Mitamura, "Effect of reducing metal on preparation of zirconium carbide powder from zirconium chloride (IV)," *Yogyo-Kyokai-Shi.*, **93**, 7–12 (1985). [in Japanese]

638. Japanese Patent 87–65, 921 (1987).

639. H. Kobayashi, M. Katoh, Y. Kamiyama and T. Mitamura, "Preparation of ZrB_2 fine powders using thermite method by reduction with Mg," *J. Ceram. Soc. Japan.*, **100**, 172–77 (1992). [in Japanese]

640. J. J. Kim and C. H. McMurtry, "TiB_2 powder production for engineered ceramics," *Ceram. Eng. Sci. Proc.*, **6**, 1313–20 (1985).

641. W. Weimin, F. Zhengyi, J. Mingji and Y. Runzhang, "Fabrication of TiB_2 powder by SHS with reduction process," *J. Chinese Ceram. Soc.*, **24**, 53–57 (1996). [in Chinese]

642. J. J. Ritter, "A low-temperature chemical process for precursors to boride and carbide ceramic powders"; pp. 21–31 in *Adv. Ceram.*, **21**, *Ceramic Powder Science.* Edited by G. L. Messing, K. S. Mazdiyasni, J. W. McCauley and R. A. Haber. American Ceramic Society, Westerville, OH, 1987.

643. H. R. Baumgartner and R. A. Steiger, "Sintering and properties of titanium diboride made from powder synthesized in a plasma-arc heater," *J. Am. Ceram. Soc.*, **67**, 207–212 (1984).

644. M. Bolech, R. Metselaar, F. K. van Dijen, F. Blomer, G. de With and P. P. J. Ramaekers, "Carbothermal preparation of Si_2N_2O powder"; pp. 527–33 in *High Tech Ceramics.* Edited by P. Vincenzini. Elsevier, Amsterdam, 1987.

645. J. Sjöberg, K. Rundgren, R. Pompe and B. Larsson, "Preparation of Si_2N_2O based sintered bodies from powders made by nitridation of amorphous silica in ammonia"; pp. 535–43 in *High Tech Ceramics.* Edited by P. Vincenzini. Elsevier, Amsterdam, 1987.

646. K. H. Jack, "The crystal chemistry of the Sialons and related nitrogen ceramics"; pp. 109–128 in *Nitrogen Ceramics.* Edited by F. L. Riley, Noordhoff, Leyden, 1977.

647. M. Mostaghaci, Q. Fan, F. L. Riley, Y. Bigay and J. P. Torre, "Densification behaviour of Sialon powders derived from aluminosilicate minerals"; pp. 149–64 in *Non-Oxide Technical and Engineering Ceramics.* Edited by S. Hampshire. Elsevier, London, 1986.

648. J.-G. Lee and I. B. Cutler, "Sinterable Sialon powder by reaction of clay with carbon and nitrogen," *Am. Ceram. Soc. Bull.*, **58**, 869–71 (1979).

649. H. Yoshimatsu, M. Mitomo, H. Mihashi, S. Ohmori, T. Yabuki, "The preparation of Sialon powder from kaolinite," *Yogyo-Kyokai-Shi*, **91**, 443–49 (1983). [in Japanese]

650. Y.-W. Cho and J. A. Charles, "Synthesis of nitrogen ceramic powder by carbothermal reduction and nitridation : II, silicon aluminium oxynitride," *Mater. Sci. Technol.*, **7**, 399–407 (1991).

651. J. Zeng, Y. Miyamoto and O. Yamada, "Combustion synthesis of Sialon powders ($Si_{6-z}Al_zO_zN_{8-z}$, z = 0.3, 0.6)," *J. Am. Ceram. Soc.*, **73**, 3700–3702 (1990).

652. N. D. Corbin, "Aluminium oxynitride spinel : a review," *J. Eur. Ceram. Soc.*, **5**, 143–54 (1989).

653. M. Ish-Shalom, "Formation of aluminium oxynitride by carbothermal reduction of aluminium oxide in nitrogen," *J. Mater. Sci. Lett.*, **1**, 147–49 (1982).

654. C. Martin and B. Cales, "Synthesis and hot pressing of transparent aluminium oxynitride," *SPIE*, **1112**, 20–24 (1989).

655. U. S. Patent 4, 686, 070 (August 1987).

656. J. Zheng and B. Forslund, "Carbothermal synthesis of aluminium oxynitride (ALON) powder : influence of starting materials and synthesis parameters," *J. Eur. Ceram. Soc.*, **15**, 1087–1100 (1995).

Subject Index

(Common precursors, chemicals etc., repeatedly mentioned through the text, are not listed here. Italicized page numbers indicate sections in which techniques of powder preparation have been discussed.)

GPSR Compliance

The European Union's (EU) General Product Safety Regulation (GPSR) is a set of rules that requires consumer products to be safe and our obligations to ensure this.

If you have any concerns about our products, you can contact us on ProductSafety@springernature.com

In case Publisher is established outside the EU, the EU authorized representative is:

Springer Nature Customer Service Center GmbH
Europaplatz 3
69115 Heidelberg, Germany

Batch number: 10755292

Printed by Printforce, the Netherlands